T0233261

A Student's Guide to the Navier–Stokes Equations

The Navier–Stokes equations describe the motion of fluids and are an invaluable addition to the toolbox of every physicist, applied mathematician, and engineer. The equations arise from applying Newton's laws of motion to a moving fluid and are considered, when used in combination with mass and energy conservation rules, to be the fundamental governing equations of fluid motion. They are relevant across many disciplines, from astrophysics and oceanic sciences to aerospace engineering and materials science. This *Student's Guide* provides a clear and focused presentation of the derivation, significance, and applications of the Navier–Stokes equations, along with the associated continuity and energy equations. Designed as a useful supplementary resource for undergraduate and graduate students, each chapter concludes with a selection of exercises intended to reinforce and extend important concepts. Video podcasts demonstrating the solutions in full are provided online, along with written solutions and other additional resources.

JUSTIN W. GARVIN is Associate Professor of Instruction in the Department of Mechanical Engineering at the University of Iowa. He has previously worked as a research engineer at Iowa's IIHR-Hydroscience and Engineering research lab and at the US Air Force Research Laboratory. His primary areas of interest are heat, fluid mechanics, and thermal physics.

Other books in the Student's Guide series:

A Student's Guide to the Navier–Stokes Equations

JUSTIN W. GARVIN
University of Iowa

Shaftesbury Road, Cambridge CB2 8EA, United Kingdom

One Liberty Plaza, 20th Floor, New York, NY 10006, USA

477 Williamstown Road, Port Melbourne, VIC 3207, Australia

314–321, 3rd Floor, Plot 3, Splendor Forum, Jasola District Centre,
New Delhi – 110025, India

103 Penang Road, #05–06/07, Visioncrest Commercial, Singapore 238467

Cambridge University Press is part of Cambridge University Press & Assessment,
a department of the University of Cambridge.

We share the University's mission to contribute to society through the pursuit of
education, learning and research at the highest international levels of excellence.

www.cambridge.org
Information on this title: www.cambridge.org/highereducation/isbn/9781009236157

DOI: 10.1017/9781009236119

© Justin W. Garvin 2023

First published 2023

A catalogue record for this publication is available from the British Library.

*A Cataloging-in-Publication data record for this book is available from the Library of
Congress*

ISBN 978-1-009-23615-7 Hardback
ISBN 978-1-009-23616-4 Paperback

Additional resources for this publication at www.cambridge.org/garvin.

About this book

This edition of *A Student's Guide to the Navier–Stokes Equations* is supported by an extensive range of digital resources, available via the book's website. These resources have been designed to support your learning and bring the textbook to life, supporting active learning and providing you with feedback.

Please visit www.cambridge.org/garvin to access this extra content.

We may update our Site from time to time and may change or remove the content at any time. We do not guarantee that our Site, or any part of it, will always be available or be uninterrupted or error free. Access to our Site is permitted on a temporary and "as is" basis. We may suspend or change all or any part of our Site without notice. We will not be liable to you if for any reason our Site or the content is unavailable at any time, or for any period.

Contents

Preface

The topic of this book deals with some of the most useful, yet complicated, equations that govern a considerable amount of our natural world. These equations are the equations of fluid motion (both liquids and gases), with the famed Navier–Stokes equations being front and center. The equations show up in a variety of areas of physics and engineering. These areas include astrophysics, condensed matter physics, materials science, geophysics, meteorology, oceanic sciences, biological sciences, as well as mechanical, aerospace, chemical, and civil engineering. In addition, the Navier–Stokes equations are also used to simulate fluids in visual effects models for movies as well as in video games.

The main goal of this book is to provide you, the student, with a friendly, but still comprehensive, path to understanding the Navier–Stokes equations as well as the mass continuity equation and energy equation (which are often seen alongside the Navier–Stokes equations). These equations as a whole make up what is commonly called the governing equations of fluid motion. This book differs from standard texts on fluid mechanics in that it focuses almost entirely on the governing equations with minimal distractions from other, albeit important, aspects of fluid mechanics. Using a somewhat conversational writing style, the book spends a lot of time discussing the meaning of the various terms of the equations. In addition, care has been taken to ensure that steps are not skipped when deriving the equations or when discussing the concepts associated with its various terms. The last chapter of the book introduces the concept of nondimensionalization (i.e., scaling) and how scaling the governing equations can reduce the number of parameters involved in a problem as well as provide insight into the physics of a problem. The examples in the book will focus mainly on what are called incompressible flows, however, there will still be a fair amount of information regarding the version of the equations that pertain to compressible flows. Along with the book, you can take advantage of video podcasts showing the solutions to the end of the chapter problems as well as some additional written material that did not make it into the final version of the book, all of which are available on the book's website.

The book is generally geared toward fourth-year students in physics and engineering. As such, in order to get the most out of the book, having some background in differential equations as well as the calculus of vectors would be helpful, if not necessary. The book would also likely be of great use if you are a beginning graduate student in science or engineering whose chosen field of study is heavily reliant on fluid mechanics. In particular, the last chapter on nondimensionalization might come in handy for those engaged in fluid mechanics research activities. If you are a mathematics student studying the mathematical side of the Navier–Stokes equations, you may find the description of the equations on a more physical basis to be insightful. In addition, if you are a lifelong learner who wants to know more about the equations that describe the motion of everyday fluids, then I hope you will find this book to be a useful addition to your self-study.

Whatever your motivation is for learning the Navier–Stokes equations, I wish you the best of luck in your quest in understanding some of the most amazing equations of nature.

Acknowledgments

When I first embarked on the journey to write a book for Cambridge's Student's Guides series, I was motivated by a little book I read on the Maxwell's equations by Professor Dan Fleisch. The book was called *A Student's Guide to Maxwell's Equations*. Even though the material was nothing new to me, I was blown away by how seemingly effortless the book flowed. It motivated me to try to write a similar book for the Navier–Stokes equations (key word here is try, as the Maxwell's equations book is a very high bar). I contacted Professor Fleisch, who turned out to be the series editor, to see if he could direct me on the steps needed to go about such a project. He put me in contact with Dr. Nicholas Gibbons, senior commissioning editor in physics, and the rest is history. I owe a great deal of thanks to these two gentlemen for their patience and their thoughtful guidance.

Additional thanks goes to Sarah Armstrong and the staff at Cambridge University Press. They are fantastic. A thank you also goes to Dr. John Mousel for reading over parts of the manuscript and providing feedback. I would also like to thank my students, in particular Colin Fisher, who have provided me with encouragement in pursuing this goal.

Finally, I would like to thank my wonderful wife, Mona Garvin, for her insightful comments and sharp eye on areas of confusion in the text.

1

Mass Conservation and the Continuity Equation

The range of applications of fluid mechanics in the natural world is vast. In particular, fluid mechanics shows up in atmospheric physics, geophysics, astrophysics, condensed matter physics, and biological physics, in addition to various engineering disciplines. As amazing as the range of applications is in fluid mechanics, most of the principles involved are the basic principles learned in any first-year physics course. However, when these principles are applied to a moving fluid, the result will be a very complicated set of equations that govern the physics of fluid motion. These equations are known as the **governing equations of fluid motion**, with the famous **Navier–Stokes equations** being front and center. This book will guide you through the details of these equations.

1.1 Conservation in Fluid Mechanics

The equations of fluid mechanics are based, in large part, on conservation principles, particularly the conservation of mass, momentum (which, as we shall see, is a consequence of Newton's laws of motion), and energy. The Navier–Stokes equations come directly from applying momentum conservation (typically in the form of Newton's second law of motion, that is, think *force = mass* times *acceleration*[1]) to a moving fluid. Although the Navier–Stokes equations are the most important of the governing equations, they are rarely (if ever) seen in isolation. Instead, they are accompanied by a conservation of mass equation (called the continuity equation or the mass continuity equation) and oftentimes alongside a conservation of energy equation (called the energy equation). The continuity equation is obtained by applying mass conservation to a moving fluid

---

[1] Note, when we formally discuss the Navier–Stokes equations, we will end up using a more general form of Newton's second law; namely that force is equal to the time derivative of momentum.

Table 1.1 *Conservation principles of fluid mechanics resulting in the governing equations*

Conservation principle[a]	Formal name of principle	Name of resulting governing fluid equation
Mass conservation	Mass conservation	Continuity equation[b]
Momentum conservation	Newton's laws of motion	Navier–Stokes equations[c]
Energy conservation	First law of thermodynamics	Energy equation

[a] Other laws, such as conservation of angular momentum and the second law of thermodynamics (which is not a conservation law), are also very important in fluid mechanics and, as we will see, are "built-in" to the three main equations.
[b] The continuity equation is sometimes used more generally as an equation describing the transport (i.e., movement) of a conserved quantity. In fluid mechanics, however, it is almost always used to describe mass conservation.
[c] Also sometimes called the momentum equations in fluid mechanics.

and the energy equation is obtained by applying energy conservation (in the form of the first law of thermodynamics) to a moving fluid. Table 1.1 summarizes the conservation principles and the name of the corresponding governing equation when each principle is applied to a moving fluid. Incidentally, historically the Navier–Stokes equations refer to the conservation of momentum when applied to a fluid (in particular, as we will discuss, a Newtonian fluid). However, some literature may refer to the whole set of equations tabulated in Table 1.1 as the Navier–Stokes equations. In this book, we will still only refer to momentum conservation as being the Navier–Stokes equations. We are not going to introduce what the equations look like up front as that could be somewhat intimidating. However, if you would like to get a sneak peek as to what the equations look like, you can check out Section 5.4.

Other principles are also important in fluid mechanics in addition to the ones mentioned above, namely the second law of thermodynamics and the conservation of angular momentum. However, these two principles are not always included as separate equations. Instead, as we shall see, they are typically "built-in" to the Navier–Stokes equations and energy equation. Also, note there is an underlying assumption with the governing equations that the fluid of interest is considered to be a continuous medium. In other words, the atomic nature of matter is not considered in the equations and, as a result, the equations start to have trouble making predictions once individual atomic interactions become an important factor. This places fluid mechanics under the umbrella of the more general subject of continuum mechanics. To give a more concrete idea of where such atomic interactions might become important, results using the governing equations to solve for airflow at standard temperature and pressure conditions in channels less than one micron in diameter (which is about a hundred times smaller than the thickness of human hair) might start to deviate from the results obtained in experiments because individual atomic interactions begin to play a role. It should still be stated, however, that even though individual atomic interactions are not accounted for in the governing equations, there are aspects of the equations that can be thought of as modeling atomic interactions in an averaged sense (which we will discuss when we reach the topic of diffusion).

Considering that the Navier–Stokes equations and the other governing equations of fluid motion rely so heavily on conservation principles, we should discuss what we mean by a conservation principle. The main idea behind a conservation principle can be stated, in words, as follows:

$$
\begin{array}{c}
\text{the amount} \\
\text{of a quantity} \\
\text{contained in} \\
\text{a system}
\end{array}
=
\begin{array}{c}
\text{amount of} \\
\text{that quantity} \\
\text{transferred to} \\
\text{the system}
\end{array}
-
\begin{array}{c}
\text{amount of} \\
\text{that quantity} \\
\text{transferred out} \\
\text{of the system}
\end{array}
+
\begin{array}{c}
\text{source (+)} \\
\text{or sink (–)} \\
\text{inside} \\
\text{the system}
\end{array}
$$

The word "system" is a term adopted from the subject of thermodynamics. **System** simply means the body of interest. The quantity in question can be anything that is conserved; examples include mass, energy, momentum, angular momentum, and electric charge.

In fluid mechanics, we typically think of conservation laws on a per time basis (or time rate basis). Thus, re-stating the idea of a conservation law on a time rate basis would look like the following:

$$
\begin{array}{c}
\text{the change} \\
\text{of a quantity} \\
\text{in a system} \\
\text{in a given} \\
\text{unit of time}
\end{array}
=
\begin{array}{c}
\text{time rate of} \\
\text{that quantity} \\
\text{entering the} \\
\text{system}
\end{array}
-
\begin{array}{c}
\text{time rate of} \\
\text{that quantity} \\
\text{leaving the} \\
\text{system}
\end{array}
+
\begin{array}{c}
\text{time rate of} \\
\text{a source or} \\
\text{sink inside} \\
\text{the system}
\end{array}
$$

A very simple example that may help you understand the conservation principle is to consider the mass of water going in and out of a bathtub. If the bathtub is our system and the quantity of interest is mass, our conservation principle for mass in the bathtub will look like this:

$$
\begin{array}{c}
\text{the change} \\
\text{of mass of water} \\
\text{in the bathtub} \\
\text{in a given} \\
\text{unit of time}
\end{array}
=
\begin{array}{c}
\text{time rate of} \\
\text{mass of water} \\
\text{entering the} \\
\text{bathtub}
\end{array}
-
\begin{array}{c}
\text{time rate of} \\
\text{mass of water} \\
\text{leaving the} \\
\text{bathtub}
\end{array}
$$

Notice how we did not bother with any source (or sink) of mass in the bathtub since water does not just appear (or disappear) from within the tub. We can now put some numbers to these terms. Suppose you are filling the bathtub with water from a faucet and the mass is entering the tub at a rate of 0.25 kilograms per second (kg/s). Now suppose you forgot to plug the drain, and the tub is losing water at 0.1 kg/s. The increase in the mass of water in the bathtub per second can be calculated as:

$$\begin{array}{c}\text{the change}\\ \text{of mass of water}\\ \text{in the bathtub}\\ \text{in a given}\\ \text{unit of time}\end{array} = 0.25 \text{ kg/s} - 0.1 \text{ kg/s} = 0.15 \text{ kg/s}$$

If the bathtub can hold 150 kg of water, then it will take about $\frac{150 \text{ kg}}{0.15 \text{ kg/s}} =$ 1000 s, roughly 17 minutes to fill up.

The rest of the chapter will continue to focus on mass conservation. This will result in the development of the continuity equation.

1.2 Conservation of Mass in One Dimension

To start our more formal discussion of mass conservation, consider the schematic in Figure 1.1. Figure 1.1 illustrates a flow going from left to right (in the x-direction). The flow could be from anything at this point: river, ocean, blood, bathtub, water in a hose, and so forth. Suppose we want to come up with an expression for mass conservation for this flow. To come up with such an expression, let us just focus on a little segment of the flow and do some "accounting" principles for mass. The little segment we are going to focus on is the box-shaped system in the middle of the flow, as shown in Figure 1.1. This box system is assumed to be a region fixed in space. In fluid mechanics, systems that are regions of a flow (often fixed in space) that allow fluid to enter and leave are sometimes given the name **control volume**. As shown in Figure 1.1, we are going to define a time rate of mass coming in from the left as $\dot{m}_{in}$ and the time rate of mass going out on the right as $\dot{m}_{out}$. Note that the units of these "mass flow rates" are, in SI units, kilograms per second. Thus, from the conservation of mass, we have:

$$\frac{dm_{sys}}{dt} = \dot{m}_{in} - \dot{m}_{out}, \tag{1.1}$$

where m_{sys} is the mass of fluid contained in the system (control volume). Equation 1.1 is nothing but our conservation principle that was discussed in the

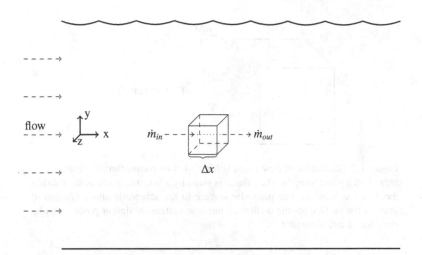

Figure 1.1 Mass passing through a box system fixed in space (control volume). The mass flow rate in ($\dot{m}_{in}$) minus the mass flow rate out ($\dot{m}_{out}$) is equal to the time rate of change of the mass in the system ($\frac{dm_{sys}}{dt}$).

previous section except now we are using variables instead of writing out the conservation principles in words.

Notice that the derivative on the left-hand side is used to denote the change of the mass of the fluid in the control volume with respect to time (t), whereas the dots ($\cdot$) above the mass flow rates denote the rate of mass entering or leaving the control volume in a given unit of time. The use of the dots is a common convention used to denote the rate of a quantity entering or leaving a system.

Now suppose we were not given mass flow rates but instead were given velocity and density going in and out of the control volume, both of which might be more easily measurable. Can we modify Equation 1.1 to be in terms of velocity and density? To start, we might want to try to find an expression for the mass flow rate in terms of velocity and density.

To determine a mass flow rate given velocity and density, consider first how you would go about determining the amount of mass passing through a surface (which we can call m_{pass}). We know that mass is a volume ($\mathcal{V}$) multiplied by the density of the fluid (ρ). The volume of fluid passing through a surface in a given amount of time (Δt) can be found by multiplying the area of the surface the fluid passes through (A) by the distance the fluid travels during that time period. The distance the fluid travels is the velocity of the fluid going into or out of the surface (which we will call u in this case[2]) multiplied by the time passed: $u\Delta t$.

---

[2] The velocity in the x-direction is, by convention in fluid mechanics, given by the variable u.

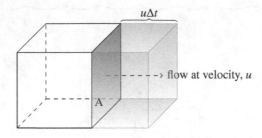

Figure 1.2 Illustration of how much volume of flow passes through a surface of area, A, in a given time, Δt. The volume is given by $u\Delta tA$, where $u\Delta t$ is the distance the flow has traveled. The particular surface in the schematic above (shaded in gray) is the surface for the outflow of our box system. A similar process can be done for an inflow surface.

A schematic of this process is given in Figure 1.2. Thus the amount of mass passing through the particular surface in Figure 1.2 is:

$$m_{pass} = \rho u\Delta tA.$$

To be a little more clear, we can make some comments on the above equation:

$$m_{pass} = \overbrace{\rho}^{\text{density}} \; \overbrace{u\Delta t}^{\substack{\text{volume of flow} \\ \text{that passes a} \\ \text{surface in time, } \Delta t}} \underbrace{A}_{\substack{\text{distance fluid} \\ \text{travels in } \Delta t}} .$$

Dividing by Δt gives a simple expression for the time rate of mass (mass flow rate) passing through a surface. Defining $\dot{m}$ as a mass flow rate then gives us the following expression:

$$\dot{m} = \frac{m_{pass}}{\Delta t} = \rho uA. \tag{1.2}$$

The product, ρu, is called the mass flux, or more specifically in this scenario, the mass flux in the x-direction.

It is important to recognize that this expression is not a general expression for the mass flow rate. For one thing, it assumes that the velocity of the flow is in a direction such that the flow comes straight out of the surface, A, and not at an angle. In addition, the velocity is inherently assumed to be uniform throughout the surface it is passing. We will come back to a more general form of mass flow rate. For now, we can just use Equation 1.2 in our quest to arrive at an equation for the mass balance in terms of density and velocity.

Plugging Equation 1.2 into Equation 1.1 yields:

$$\frac{dm_{sys}}{dt} = (\rho u A)_{in} - (\rho u A)_{out}. \tag{1.3}$$

Focusing now on the left side of Equation 1.3, we can write m_{sys} as the density of the fluid multiplied by the volume of our box system ($A\Delta x$), where Δx is the length of the box system in the x-direction. This would yield:

$$\frac{d}{dt}(\rho A \Delta x) = (\rho u A)_{in} - (\rho u A)_{out}.$$

Pulling out the $A\Delta x$ from the time derivative (since it is a constant) and dividing it out leads to:

$$\frac{\partial \rho}{\partial t} = \frac{(\rho u)_{in} - (\rho u)_{out}}{\Delta x}. \tag{1.4}$$

The reason for switching to a partial derivative on the left-hand side of Equation 1.4 will become apparent when we deal with the more general case later in the chapter. For now, just recognize that the density is dependent not only on time but also on space (i.e., is also a function of x), hence the partial derivative. We can rearrange Equation 1.4 to get all terms on the left of the equal sign:

$$\frac{\partial \rho}{\partial t} + \frac{(\rho u)_{out} - (\rho u)_{in}}{\Delta x} = 0.$$

Recognize that the second term on the left-hand side of the above equation is just the change in ρu over the change in x. This is the slope of ρu in the x-direction. Taking the limit of this slope as $\Delta x \to 0$ makes the second term become a derivative, that is:

$$\frac{\partial \rho}{\partial t} + \frac{\partial (\rho u)}{\partial x} = 0. \tag{1.5}$$

Equation 1.5 is a one-dimensional version of the continuity equation in Cartesian coordinates. It states, in effect, that the change of density (or mass per volume) with respect to time plus the spatial change of the mass flux (ρu) is equal to zero. It comes directly from a simple mass conservation equation, Equation 1.1. As you can see, even in the one-dimensional case, this equation is a complicated differential equation (i.e., an equation that includes derivatives). In particular, it is what is known as a partial differential equation (PDE) because the dependent variable (namely density, ρ) is a function of both time (t) and space (x). Thus, the derivatives seen in the differential equation are partial derivatives. As we will see, the governing equations are often written as partial differential equations. Unfortunately, PDEs are quite difficult to solve, and thus many times the solution of them requires the use of a computer. We will discuss this further at various times as we proceed.

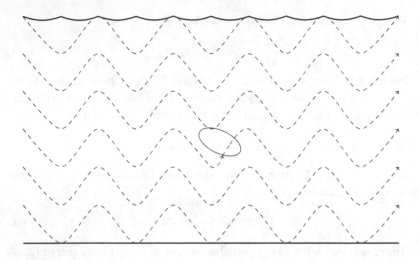

Figure 1.3 Flow passing through a control volume shaped as an ellipsoid.

We have looked at a simple one-dimensional situation. We can now extend the idea to obtain a general continuity equation for multiple dimensions.

1.3 The Continuity Equation

In this section, we are going to take a look at the continuity equation in a more general form. Instead of a 1-D flow situation that was studied earlier, consider the control volume in the middle of the wavy flow as shown in Figure 1.3. The shape of the control volume is arbitrary. Sometimes in fluid mechanics and continuum mechanics texts, such arbitrary volumes are called potatoes and are thus sometimes shaped as such. The shape we will use for our control volume will just be an ellipsoid. Suppose flow is passing through our control volume. We can calculate the change in mass contained in our control volume with respect to time (i.e., the left-hand side of Equation 1.1) by taking the time derivative of the mass of fluid contained in the fixed control volume. The mass of the fluid inside the control volume (m_{sys}), is given by the volume integral of density (ρ):

$$m_{sys} = \iiint_{\mathcal{V}} \rho d\mathcal{V}. \tag{1.6}$$

You might be wondering why mass is not just the density multiplied by the volume of our ellipsoid ($\mathcal{V}$). The reason for the volume integral is that we are now going to consider that the density may vary within the control volume.

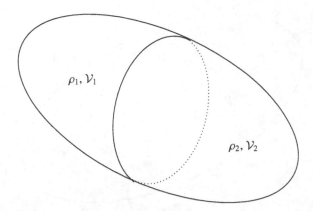

Figure 1.4 The control volume split into two pieces, each with its own volume and density. The total mass of the system in this case is given by: $m_{sys} = \rho_1 \mathcal{V}_1 + \rho_2 \mathcal{V}_2$. If we were to break up the system into infinitesimal volumes, each with a volume of $d\mathcal{V}$, the total mass would then be an infinite summation (an integral) of the $\rho d\mathcal{V}$s over the whole volume, that is, $\iiint_\mathcal{V} \rho d\mathcal{V}$.

Therefore, the integral takes into account a varying density within the volume. Thus, if the system is broken into two pieces, as shown in Figure 1.4, where one piece has a density of ρ_1 and a volume of $\mathcal{V}_1$ and the other piece has a density of ρ_2 and a volume of $\mathcal{V}_2$, then the mass of the system will just be:

$$m_{sys} = \rho_1 \mathcal{V}_1 + \rho_2 \mathcal{V}_2.$$

Since our control volume could theoretically be broken up into an infinite number of infinitesimal "pieces," all with a separate density and infinitesimal volume $(d\mathcal{V})$, an integral is used to "sum" over all of the little masses (i.e., $\rho d\mathcal{V}$) to get the total mass of the system, m_{sys}.

Next, we need to determine the rate of mass coming in or going out of the control volume. Unlike our 1-D version where the velocity of the flow was only in the x-direction, this time the velocity of the flow coming out of the surface might be at an angle and not necessarily coming "straight out" of the surface. We will denote the velocity of the flow by the vector $\vec{V}$. Generally, in Cartesian coordinates, the velocity vector can be broken up into components (with $\hat{i}, \hat{j}$, and $\hat{k}$ being the base vectors in the x-, y-, and z-directions, respectively) via:

$$\vec{V} = u\hat{i} + v\hat{j} + w\hat{k} \tag{1.7}$$

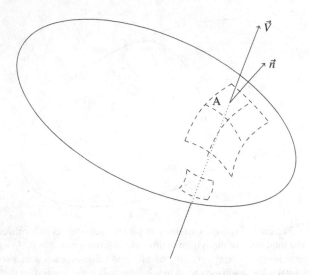

Figure 1.5 The ellipsoid control volume broken up into area segments (only a few area segments are shown). The velocity vector of the flow ($\vec{V}$) coming out of one of the area segments is shown, as is the unit normal ($\vec{n}$). The surface area (A) is also provided.

Or in column vector form:

$$\vec{V} = \begin{pmatrix} u \\ v \\ w \end{pmatrix} \tag{1.8}$$

where u is the x-component of velocity, v is the y-component of velocity, and w is the z-component of velocity. This naming convention for the velocity components might seem odd but is very standard in fluid dynamics texts for Cartesian coordinates.

The surface of our control volume can be broken up into little pieces (area segments), each one having a unique velocity vector denoting the velocity of the flow going in or out. Figure 1.5 provides an illustration. There are a few area segments shown, with one in particular that is marked with a surface area of A. The area should be small enough such that we can define what is called an outward unit normal (or unit normal or outward normal) for that given surface, $\vec{n}$. The unit normal is a vector of unit length that is perpendicular, or normal, to the area segment and points away (hence outward) from the volume. In Cartesian coordinates, the unit normal can be written as:

$$\vec{n} = n_x \hat{i} + ny\hat{j} + n_z \hat{k} \tag{1.9}$$

or in column vector form:

$$\vec{n} = \begin{pmatrix} n_x \\ n_y \\ n_z \end{pmatrix}, \tag{1.10}$$

where the components in x-, y-, and z-coordinates are given by n_x, n_y, and n_z, respectively.

Using the velocity vector and unit normal for a given surface, we can calculate the mass flow rate going through an area, A, via:

$$\dot{m} = \rho \vec{V} \cdot \vec{n} A. \tag{1.11}$$

This is very similar to the 1-D situation except now, instead of u for velocity, we are using $\vec{V} \cdot \vec{n}$, which is the component of velocity in the normal direction. The $\vec{V} \cdot \vec{n}$ is the dot product of the velocity ($\vec{V}$) with the unit normal to the surface ($\vec{n}$). You may recall from calculus that the dot product "projects" one vector onto another. As a reminder, the dot product of two general vectors, $\vec{A}$ and $\vec{B}$, at an angle θ relative to each other, have the following relationship:

$$\vec{A} \cdot \vec{B} = |\vec{A}||\vec{B}| \cos(\theta)$$

where the $|\ \ |$ indicate magnitudes of the vectors. If we replace $\vec{A}$ with $\vec{V}$ and $\vec{B}$ with $\vec{n}$, we get:

$$\vec{V} \cdot \vec{n} = |\vec{V}||\vec{n}| \cos(\theta). \tag{1.12}$$

Since $|\vec{n}|$ is equal to 1 because it has unit length, we have $\vec{V} \cdot \vec{n} = |\vec{V}| \cos(\theta)$, which is the component of the velocity in the normal direction (V_n). An illustration of this projection idea is given in Figure 1.6. This means that only the normal component of velocity (i.e., the velocity coming straight out of the surface) contributes to the mass flow rate. In our earlier section where we determined the mass flow rate to be $\rho u A$, the x-velocity (u) was coming straight out of (or straight into) the right and left surfaces of our control volume. Here, the u is replaced with $\vec{V} \cdot \vec{n}$.

Calculating the dot product, if you are given the components of velocity and outward normal in Cartesian coordinates, is a fairly straightforward calculation. You simply multiply the corresponding components of the velocity and unit normal and add them up[3]:

$$\vec{V} \cdot \vec{n} = (u\hat{i} + v\hat{j} + w\hat{k}) \cdot (n_x\hat{i} + n_y\hat{j} + n_z\hat{k})$$

$$= un_x + vn_y + wn_z.$$

---

[3] Note, the dot product calculation in non-orthogonal coordinates would require the involvement of what is called a metric tensor.

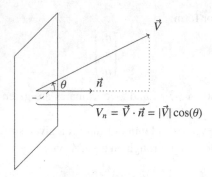

Figure 1.6 The calculation of the normal velocity component (i.e., the velocity coming out straight out of a surface).

Another useful calculation approach to take when calculating the dot product for Cartesian coordinates is to matrix multiply a row vector with a column vector via:

$$\vec{V} \cdot \vec{n} = \vec{V}^{\dagger} \vec{n} = \begin{pmatrix} u & v & w \end{pmatrix} \begin{pmatrix} n_x \\ n_y \\ n_z \end{pmatrix} = un_x + vn_y + wn_z$$

where the $\dagger$ indicates taking the transpose (i.e., flip the columns to rows and visa versa) of the velocity vector.

Continuing with our discussion of mass flux, to get the total mass rate flowing in or out of an arbitrarily shaped volume we will need to add up all of the contributing mass flow rates at each area segment. This results in the total mass flow rate being the summation of the mass flow rate of each area segment, that is:

$$\dot{m}_{total} = \sum_{i=1}^{N} \rho_i \vec{V}_i \cdot \vec{n}_i A_i \qquad (1.13)$$

where N is the number of area segments. The subscript i denotes an individual segment on the control volume. If we make each area segment infinitesimal in size (i.e., equal to dA) and sum over an infinite number of them, we can replace the summation ($\sum$) with an area integral over the entire surface:

$$\dot{m}_{total} = \oiint_A \rho \vec{V} \cdot \vec{n} dA. \qquad (1.14)$$

The circle in the middle of the integral in Equation 1.14 implies integrating over an entire, closed surface. This area integral is very indicative of a flux-type integral. A flux-type integral is an area integral over a vector quantity dotted with the outward normal. This vector quantity is called the flux of a quantity, which is the time rate of that quantity per unit area (or the quantity per volume times velocity).[4] In this case, the flux is a mass flux (which we denote $\vec{m}''$) and is given by:

$$\vec{m}'' = \rho\vec{V}. \tag{1.15}$$

Equation 1.15 describes the rate of mass per area passing through a particular surface. It has units of kg/m^{2}s, which is a rate of mass (kg/s) per area. Notice it is a vector quantity since density is a scalar (i.e., has no directional association), and the velocity is a vector. Multiplying a scalar with a vector results in a vector. Thus, it has vector components in the x-, y-, and z-directions. The mass flow rate, Equation 1.14, describes the total rate of mass passing through the total area that encloses a volume.

As a general rule of thumb, the rate of a quantity, X, denoted as $\dot{X}$, passing through the surface of a volume is related to the area integral of the flux of that quantity ($\vec{X}''$) dotted with the outward normal vector ($\vec{n}$), that is:

$$\underbrace{\dot{X}}_{\substack{\text{rate of some} \\ \text{quantity, X}}} = \oiint_A \underbrace{\vec{X}''}_{\substack{\text{flux of some} \\ \text{quantity, X}}} \cdot \vec{n} dA. \tag{1.16}$$

Expressions similar to Equation 1.16 are used often in fluid mechanics to establish relationships between rates of mass, momentum, and energy with their fluxes. We will encounter equations of the form of Equation 1.16 as we proceed through the book.

The general equation for the mass flow rate, Equation 1.14, is the same as the negative of the mass flow rate going into the volume minus the mass flow rate leaving the volume, that is:

$$\dot{m}_{total} = \oiint_A \rho\vec{V} \cdot \vec{n} dA = -(\dot{m}_{in} - \dot{m}_{out}). \tag{1.17}$$

The reason that Equation 1.14 is the negative of the mass rate entering minus the mass rate leaving is because of our convention for the unit normal ($\vec{n}$) being pointed away from the control volume as opposed to inward. Since $\vec{n}$ is pointed outward, then the dot product of the mass flux vector ($\rho\vec{V}$) with $\vec{n}$ will be positive if the mass flux vector is pointing away from the control volume (i.e., mass is leaving the volume) and negative if the mass flux vector is pointing towards the body (i.e., mass is entering the volume).

Now plug Equation 1.6 $\left(m_{sys} = \iiint_V \rho d\mathcal{V}\right)$ and Equation 1.17 into the mass conservation, Equation 1.1 $\left(\frac{dm_{sys}}{dt} = \dot{m}_{in} - \dot{m}_{out}\right)$, to get:

---

[4] Depending on the field, you may see somewhat different definitions for flux.

$$\frac{d}{dt} \iiint_\mathcal{V} \rho d\mathcal{V} = - \oiint_A \rho \vec{V} \cdot \vec{n} dA. \tag{1.18}$$

The next step is to bring the time derivative into the integral on the left-hand side. Before doing so, note that the derivative on the left-hand side is an ordinary derivative. The reason for this is because the volume integral on the left-hand side "integrates out" any spatial dependence. As a simple example of what is meant by "integrating out" any spatial dependence, consider a 1D example with $\rho = xt$ where x goes from 0 to 1 meter. Thus $\iiint_\mathcal{V} \rho d\mathcal{V} = \int_{0m}^{1m} \rho dx = \int_{0m}^{1m} xt dx = \frac{1}{2}x^2 t \Big|_{x=0m}^{1m}$. The result of this integral is going to be (once the limits are taken): $\frac{1}{2}t$, which is a function of time and not x. Hence, x (or space) was integrated out. Therefore, the derivative on the left-hand side of Equation 1.18 is an ordinary derivative. In other words:

$$\frac{d}{dt} \underbrace{\iiint_\mathcal{V} \overbrace{\rho}^{\substack{\text{a function} \\ \text{of space} \\ \text{and time}}} d\mathcal{V}}_{\substack{\text{a function of time} \\ \text{(space is integrated out)}}} = - \oiint_A \rho \vec{V} \cdot \vec{n} dA.$$

Given that the volume is fixed and not changing size or moving, we can simply bring the time derivative inside the integral, thus now taking the derivative of density. Since density is a function of both time and space, we need to change the ordinary derivative to a partial derivative. Thus:

$$\iiint_\mathcal{V} \frac{\partial \rho}{\partial t} d\mathcal{V} = - \oiint_A \rho \vec{V} \cdot \vec{n} dA. \tag{1.19}$$

This is somewhat of an "informal" way of bringing the derivative inside the integral and works well in this case because the volume does not change with time. However, we will come across a similar situation in a later chapter where we will need to be a little more careful about bringing the derivative in under the integral sign.

We can bring the right-hand side of Equation 1.19 to the left-hand side to get an **integral form of the continuity equation** that is often seen:[5]

$$\boxed{\iiint_\mathcal{V} \frac{\partial \rho}{\partial t} d\mathcal{V} + \oiint_A \rho \vec{V} \cdot \vec{n} dA = 0}. \tag{1.20}$$

---

[5] Note that we derived this equation assuming a fixed control volume, i.e., one that does not move. We could have generalized this equation a little bit to include a control volume moving at some velocity, $\vec{W}$, which would have resulted in the following equation:
$\iiint_\mathcal{V} \frac{\partial \rho}{\partial t} d\mathcal{V} + \oiint_A \rho \left(\vec{V} - \vec{W} \right) \cdot \vec{n} dA = 0.$

For some situations, this would be a good ending point for the mass conservation equation. However, many times the governing equations are written as differential equations, as opposed to Equation 1.20, which is considered an integral form (even though there is still a derivative in the equation). To turn Equation 1.20 into a differential equation, it would be ideal if the integrals were either both area integrals or volume integrals. It turns out, that making the area integral on the left-hand side into a volume integral can be done if you remember the divergence theorem from vector calculus. The divergence theorem converts an area integral where the integrand is a vector quantity dotted with the outward unit normal to the volume integral of the divergence of that same vector quantity. In our case, the vector quantity is $\rho \vec{V}$. Applying the divergence theorem to the area integral in Equation 1.20 yields:

$$\oiint_A \rho \vec{V} \cdot \vec{n} dA \overset{\substack{\text{using} \\ \text{divergence} \\ \text{theorem}}}{=} \iiint_V \underbrace{\vec{\nabla} \cdot \left(\rho \vec{V} \right)}_{\text{divergence of } \rho \vec{V}} dV, \tag{1.21}$$

where the $\vec{\nabla}$ is the nabla, or del, operator. When the $\vec{\nabla}$ operator is followed by a dot ($\cdot$), the divergence of a vector is to be taken. More will be said about the del operator and the divergence theorem shortly. For now, just sit tight, for we have a volume integral for the flux relationship that we can use.

Plugging Equation 1.21 into Equation 1.20 leads to:

$$\iiint_V \frac{\partial \rho}{\partial t} dV + \iiint_V \vec{\nabla} \cdot \left(\rho \vec{V} \right) dV = 0.$$

Combining into one volume integral and factoring out the dV from the two terms gives us:

$$\iiint_V \underbrace{\left(\frac{\partial \rho}{\partial t} + \vec{\nabla} \cdot \left(\rho \vec{V} \right) \right)}_{\text{needs to be zero}} dV = 0. \tag{1.22}$$

The only way for this volume integral to be zero for any and all volumes is if the integrand is zero.[6] This brings us to the differential equation form of the continuity equation that is most typically seen:

$$\boxed{\frac{\partial \rho}{\partial t} + \vec{\nabla} \cdot \left(\rho \vec{V} \right) = 0} \tag{1.23}$$

---

[6] Note, I will sometimes say: shrink to an infinitesimal volume when converting from an integral equation to a differential equation.

Equation 1.23 is often called the **conservation form of the continuity equation**. It can also be called conservative form or Eulerian form.[7]

1.4 Discussion of the Continuity Equation

To begin our discussion of the continuity equation, let's first take a look at the $\vec{\nabla}$ operator that shows up in Equation 1.23. An operator performs some operation on an input. For example, a derivative is an operator. By itself, a derivative means nothing, it needs a function (input) to operate on. You may recall that the del operator is associated with a number of operations, such as the divergence ($\vec{\nabla} \cdot ()$), the curl ($\vec{\nabla} \times ()$), the gradient ($\vec{\nabla}()$), and the Laplacian ($\nabla^2 = \vec{\nabla} \cdot \vec{\nabla}()$). We will discuss these operations as we encounter them in the text. The del symbol many times is usually written in Cartesian coordinates to look like a vector, hence the reason for the $\rightarrow$ above the ∇ symbol, even though it is an operator and not a vector. In Cartesian coordinates, the $\vec{\nabla}$ can be written as:

$$\vec{\nabla} \equiv \hat{i}\frac{\partial}{\partial x} + \hat{j}\frac{\partial}{\partial y} + \hat{k}\frac{\partial}{\partial z}.$$

Another way the $\vec{\nabla}$ might be written in Cartesian coordinates is in row vector form:

$$\vec{\nabla} \equiv \left(\frac{\partial}{\partial x} \quad \frac{\partial}{\partial y} \quad \frac{\partial}{\partial z}\right).$$

Most of the book will involve Cartesian coordinates (when not using the $\vec{\nabla}$ notation), just because it is the easiest coordinate system to work with and the operations in Cartesian coordinates are the least likely to distract from a discussion on the physics.

The Divergence, $\vec{\nabla} \cdot ()$ When the divergence operates on a vector, the result is a scalar.

The divergence operation in Cartesian coordinates is very similar to the dot product operation in Cartesian coordinates except instead of doing matrix multiplication, you are taking derivatives. Taking, as an example, the divergence of $\rho\vec{V}$ would look something like this:

$$\vec{\nabla} \cdot \left(\rho\vec{V}\right) = \left(\hat{i}\frac{\partial}{\partial x} + \hat{j}\frac{\partial}{\partial y} + \hat{k}\frac{\partial}{\partial z}\right) \cdot \left(\rho u\hat{i} + \rho v\hat{j} + \rho w\hat{k}\right)$$

$$= \frac{\partial(\rho u)}{\partial x} + \frac{\partial(\rho v)}{\partial y} + \frac{\partial(\rho w)}{\partial z}.$$

---

[7] We will discuss what Eulerian means in Chapter 2.

Note that the derivative associated with each base vector in the divergence operation is performed on the corresponding vector component. As an example, the $\frac{\partial}{\partial x}$ is "associated" with the base vector $\hat{i}$ and thus the derivative with respect to x is performed on the x-component of our vector (in this case, the x-component is ρu).

In matrix form for Cartesian coordinates, the divergence can be written as:

$$\vec{\nabla} \cdot \left(\rho \vec{V} \right) = \begin{pmatrix} \frac{\partial}{\partial x} & \frac{\partial}{\partial y} & \frac{\partial}{\partial z} \end{pmatrix} \begin{pmatrix} \rho u \\ \rho v \\ \rho w \end{pmatrix} = \frac{\partial (\rho u)}{\partial x} + \frac{\partial (\rho v)}{\partial y} + \frac{\partial (\rho w)}{\partial z}$$

where you would perform the operation much like you would matrix multiplication except that you would be applying the derivatives of the row vector to the column vector as opposed to multiplying.[8]

Thus, in Cartesian coordinates, the 3D continuity equation is written as:

$$\frac{\partial \rho}{\partial t} + \frac{\partial (\rho u)}{\partial x} + \frac{\partial (\rho v)}{\partial y} + \frac{\partial (\rho w)}{\partial z} = 0 \qquad (1.24)$$

Recall from Section 1.2 that we came up with a one dimensional version of the continuity equation, Equation 1.5. This is the same as if v and w in Equation 1.24 are zero, that is:

$$\frac{\partial \rho}{\partial t} + \frac{\partial (\rho u)}{\partial x} = 0.$$

Below we label each of the terms of the continuity equation in Cartesian coordinates:

$$\underbrace{\frac{\partial \rho}{\partial t}}_{\substack{\text{change of mass} \\ \text{in the system} \\ \text{per unit time}}} + \underbrace{\frac{\partial \overbrace{(\rho u)}^{\substack{\text{mass flux} \\ \text{in } x}}}{\partial x}}_{\substack{\text{total time} \\ \text{rate of mass} \\ \text{leaving/entering system} \\ \text{in the } x\text{-direction}}} + \underbrace{\frac{\partial \overbrace{(\rho v)}^{\substack{\text{mass flux} \\ \text{in } y}}}{\partial y}}_{\substack{\text{total time} \\ \text{rate of mass} \\ \text{leaving/entering system} \\ \text{in the } y\text{-direction}}} + \underbrace{\frac{\partial \overbrace{(\rho w)}^{\substack{\text{mass flux} \\ \text{in } z}}}{\partial z}}_{\substack{\text{total time} \\ \text{rate of mass} \\ \text{leaving/entering system} \\ \text{in the } z\text{-direction}}} = 0$$

total time rate of mass leaving/entering the system
(i.e., $\dot{m}_{out} - \dot{m}_{in}$)

---

[8] A word of caution, this approach does not work in general but is very useful for Cartesian coordinates.

In general, coordinate-free (or, del-form) notation, we have:

$$\underbrace{\frac{\partial \rho}{\partial t}}_{\substack{\text{change of mass} \\ \text{in the system} \\ \text{per unit time}}} + \overbrace{\vec{\nabla} \cdot \underbrace{\left(\rho \vec{V}\right)}_{}}^{\text{divergence \ mass flux}} \underbrace{}_{\substack{\text{total time rate} \\ \text{of mass leaving/entering} \\ \text{the system} \\ \text{(i.e., } \dot{m}_{out} - \dot{m}_{in})}} = 0.$$

Keep in mind that the continuity equation in its differential form is actually in mass per volume, thus the units of each term is kg/m^{3}s as opposed to kg/s.

Looking at the integral form, we have:

$$\underbrace{\iiint_V \frac{\partial \rho}{\partial t} d\mathcal{V}}_{\substack{\text{change of mass} \\ \text{in the system} \\ \text{per unit time}}} + \underbrace{\oiint_A \overbrace{\rho \vec{V}}^{\text{mass flux}} \cdot \vec{n} dA}_{\substack{\text{total time rate} \\ \text{of mass leaving/entering} \\ \text{the system} \\ \text{(i.e., } \dot{m}_{out} - \dot{m}_{in})}} = 0$$

As you can see, each one of the equations has the same type of terms: a change of mass term (or density in the case of the differential equation version) and the time rate of mass leaving and/or entering the system (i.e., the mass flow rate).

1.4.1 The Divergence Theorem

We used a theorem, without much discussion, called the divergence theorem to convert our area integral of Equation 1.20 into a volume integral, Equation 1.22. We then used Equation 1.22 to obtain a differential equation form, known as the conservation form, of the continuity equation, Equation 1.23. In this section we are going to illustrate, using an example in Cartesian coordinates, how the divergence theorem works. The reason for going through such an exercise is that the divergence theorem will be used in the discussion of the other governing equations and is a very important theorem in fluid mechanics.

Suppose we have a Cartesian control volume with flow passing through at an arbitrary angle with a velocity of $\vec{V}$, as shown in Figure 1.7. Recall that our mass flow rate in an integral form is given by Equation 1.14, rewritten here:

$$\dot{m}_{total} = \oiint_A \rho \vec{V} \cdot \vec{n} dA.$$

Also recall that we can approximate this integral using a summation over a given number of area segments that the volume is broken up into (as seen in Equation 1.13 and rewritten here):

$$\dot{m}_{total} = \sum_{i=1}^{N} \rho_i \vec{V}_i \cdot \vec{n}_i A_i.$$

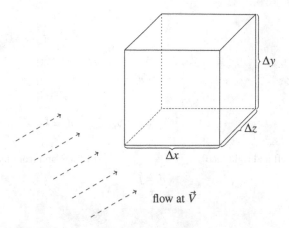

Figure 1.7 Incoming flow past a Cartesian control volume.

We can break out our approximate mass flow rate equation for our Cartesian volume by summing over all of the faces (i.e., area segments) of the cube, assuming that the density and velocity do not vary within each face. Figure 1.8 illustrates the velocity and unit normals at each of the faces of our control volume. Our approximate mass flow rate equation (using subscripts l, r, t, b, f, and bk to denote the left, right, top, bottom, front, and back sides of our cube) leads to:

$$
\begin{aligned}
\dot{m}_{total} &= \sum_{i=1}^{6} \rho_i \vec{V}_i \cdot \vec{n}_i A_i \\
&= \rho_r \vec{V}_r \cdot \vec{n}_r A_r + \rho_l \vec{V}_l \cdot \vec{n}_l A_l \\
&\quad + \rho_t \vec{V}_t \cdot \vec{n}_t A_t + \rho_b \vec{V}_b \cdot \vec{n}_b A_b \\
&\quad + \rho_f \vec{V}_f \cdot \vec{n}_f A_f + \rho_{bk} \vec{V}_{bk} \cdot \vec{n}_{bk} A_{bk}.
\end{aligned}
\tag{1.25}
$$

Now, we can define our normal vectors for each of the sides (given in Figure 1.8) to be (with the z-component coming out of the page):

$$
\vec{n}_l = -\hat{i} \quad \vec{n}_r = \hat{i} \quad \vec{n}_b = -\hat{j} \quad \vec{n}_t = \hat{j} \quad \vec{n}_{bk} = -\hat{k} \quad \vec{n}_f = \hat{k}.
\tag{1.26}
$$

Plugging in $\vec{V} = u\hat{i} + v\hat{j} + w\hat{k}$ and the definition of our normal vectors from Equation 1.26 into Equation 1.25, we get:

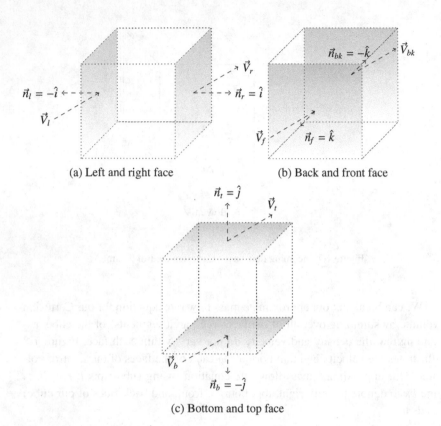

(a) Left and right face   (b) Back and front face

(c) Bottom and top face

Figure 1.8 The flow past the various surfaces of a Cartesian control volume.

$$\dot{m}_{total} = \sum_{i=1}^{6} \rho_i \vec{V}_i \cdot \vec{n}_i A_i$$

$$= \rho_r \left(u_r \hat{i} + v_r \hat{j} + w_r \hat{k} \right) \cdot \left(\hat{i} \right) A_r + \rho_l \left(u_l \hat{i} + v_l \hat{j} + w_l \hat{k} \right) \cdot \left(-\hat{i} \right) A_l$$
$$+ \rho_t \left(u_t \hat{i} + v_t \hat{j} + w_t \hat{k} \right) \cdot \left(\hat{j} \right) A_t + \rho_b \left(u_b \hat{i} + v_b \hat{j} + w_b \hat{k} \right) \cdot \left(-\hat{j} \right) A_b$$
$$+ \rho_f \left(u_f \hat{i} + v_f \hat{j} + w_f \hat{k} \right) \cdot \left(\hat{k} \right) A_f + \rho_{bk} \left(u_{bk} \hat{i} + v_{bk} \hat{j} + w_{bk} \hat{k} \right) \cdot \left(-\hat{k} \right) A_{bk}$$
$$(1.27)$$

Performing the dot products in Equation 1.27 leads to a simplified form:

$$\dot{m}_{total} = \sum_{i=1}^{6} \rho_i \vec{V}_i \cdot \vec{n}_i A_i$$
$$= \rho_r u_r A_r - \rho_l u_l A_l + \rho_t v_t A_t - \rho_b v_b A_b + \rho_f w_f A_f - \rho_{bk} w_{bk} A_{bk}.$$
$$(1.28)$$

The values of the surface areas are nothing but the areas of each face:

$$A_r = A_l = \Delta y \Delta z$$
$$A_t = A_b = \Delta x \Delta z \tag{1.29}$$
$$A_f = A_{bk} = \Delta x \Delta y.$$

We can plug Equation 1.29 into Equation 1.28 to get:

$$\dot{m}_{total} = \sum_{i=1}^{6} \rho_i \vec{V}_i \cdot \vec{n}_i A_i$$
$$= \rho_r u_r \Delta y \Delta z - \rho_l u_l \Delta y \Delta z + \rho_t v_t \Delta x \Delta z - \rho_b v_b \Delta x \Delta z \tag{1.30}$$
$$+ \rho_f w_f \Delta x \Delta y - \rho_{bk} w_{bk} \Delta x \Delta y.$$

We can now pull a trick. The trick we can pull is to factor out the volume from the whole right hand side of Equation 1.30. That is, we can factor out $\Delta x \Delta y \Delta z$ to get:

$$\dot{m}_{total} = \sum_{i=1}^{6} \rho_i \vec{V}_i \cdot \vec{n}_i A_i$$
$$= \left(\frac{\rho_r u_r - \rho_l u_l}{\Delta x} + \frac{\rho_t v_t - \rho_b v_b}{\Delta y} + \frac{\rho_f w_f - \rho_{bk} w_{bk}}{\Delta z} \right) \Delta x \Delta y \Delta z. \tag{1.31}$$

Taking the limits as $\Delta x \to 0$, $\Delta y \to 0$, $\Delta z \to 0$ changes the slopes to derivatives. In addition, we can denote $\Delta x \Delta y \Delta z$ as the volume of an infinitesimal Cartesian element $(d\mathcal{V})$. Denoting the total mass flow rate of this infinitesimally sized volume as $\delta \dot{m}_{total}$ leaves us with:

$$\delta \dot{m}_{total} = \left(\underbrace{\frac{\rho_r u_r - \rho_l u_l}{\Delta x}}_{\frac{\partial(\rho u)}{\partial x}} + \underbrace{\frac{\rho_t v_t - \rho_b v_b}{\Delta y}}_{\frac{\partial(\rho v)}{\partial y}} + \underbrace{\frac{\rho_f w_f - \rho_{bk} w_{bk}}{\Delta z}}_{\frac{\partial(\rho w)}{\partial z}} \right) \underbrace{\Delta x \Delta y \Delta z}_{d\mathcal{V}}. \tag{1.32}$$
$$\underbrace{}_{\vec{\nabla} \cdot (\rho \vec{V})}$$

The right-hand side of Equation 1.32 is the divergence of the mass flux $(\rho \vec{V})$ multiplied by the volume of the Cartesian element. Therefore:

$$\delta \dot{m}_{total} = \vec{\nabla} \cdot \left(\rho \vec{V} \right) d\mathcal{V}. \tag{1.33}$$

Equation 1.33 is a general expression for the relationship between the infinitesimal mass flow rate and an infinitesimal control volume. We can get a general result for a finite sized control volume by integrating both sides:

$$\dot{m}_{total} = \iiint_{\mathcal{V}} \vec{\nabla} \cdot \left(\rho \vec{V} \right) d\mathcal{V}. \tag{1.34}$$

Note that integrating $\delta \dot{m}_{total}$ gives just $\dot{m}_{total}$ as opposed to $\dot{m}_{total}$ evaluated at some final state minus an initial state. The reason is that there is no such thing as a final and initial state for the mass flow rate. This is why we wrote the infinitesimal mass flow rate as $\delta \dot{m}_{total}$ instead of $d\dot{m}_{total}$. Ultimately, this boils down to the distinction between an exact differential and an inexact differential. We are not going to harp on this distinction, although this idea of exact differentials versus inexact differentials does come up in fluid dynamics and, in particular, thermodynamics.

We can now equate Equation 1.34 with Equation 1.14 (i.e., $\dot{m}_{total} = \oiint_{A} \rho \vec{V} \cdot \vec{n} dA$) to get:

$$\oiint_{A} \rho \vec{V} \cdot \vec{n} dA = \iiint_{\mathcal{V}} \vec{\nabla} \cdot \left(\rho \vec{V} \right) d\mathcal{V}. \tag{1.35}$$

We see that Equation 1.35 has changed an area integral over an entire surface of the volume to a volume integral. This is the essence of the divergence theorem. Equation 1.35 was using the mass flux vector $(\rho \vec{V})$ as the "input" vector to the theorem, but this can be replaced with a general vector, $\vec{f}$. Thus, the **divergence theorem for a vector quantity, $\vec{f}$, passing through a volume, $\mathcal{V}$, is given by the following relationship:**

$$\boxed{\oiint_{A} \vec{f} \cdot \vec{n} dA = \iiint_{\mathcal{V}} \vec{\nabla} \cdot \vec{f} d\mathcal{V}} \tag{1.36}$$

1.4.2 The Unknowns in the Continuity Equation

The governing equations of fluid mechanics are field equations, meaning that their unknowns (which we may call field variables or flow variables) have values at particular points in space as well as time. As we will see, the unknowns for the governing equations will be velocity, pressure, and, in some cases, temperature and density. You have already been exposed to velocity and density via the continuity equation, but in some ways, the most familiar field variable might be temperature. Temperature is very often seen on weather maps. A typical weather map of temperature might color the temperature field over a region of space (by convention, the hotter areas are usually red while cooler areas are typically shaded in blue). In other situations, a weather map of a particular region might plot vectors of wind (i.e., the velocity vector field, $\vec{V}$) over a particular geographical region. Similarly, continuing with our example of weather, pressure could also be plotted to show pressure highs and lows in a region.

In this chapter, our focus is the continuity equation, which in differential form is:

$$\frac{\partial \rho}{\partial t} + \vec{\nabla} \cdot \left(\rho \vec{V}\right) = 0$$

We have two main variables that show up predominately, the density (ρ) and the flow velocity ($\vec{V}$). The independent variables are time (t) and space (e.g., x, y, and z). Thus, it appears that velocity and density are our two unknowns (actually, we have four unknowns because velocity has three components, one for each spatial direction, since it is a vector quantity). This begs the question, is the continuity equation predominately an equation for velocity or density? We will find that the velocity field will be determined by the all-important (and the book's namesake) Navier–Stokes equations. Therefore, the continuity equation is considered an equation for density. Density is considered the "dominant" term in the continuity equation. Meaning that it not only shows up in every term of the continuity equation, there is also a time-derivative of density. Hence, the continuity equation deals with the time evolution of density, i.e., determines how the density changes as the system progresses in time. The velocity field, in some ways, acts as an input to the continuity equation. We will examine the inputs and unknowns of the governing equations as we progress through this book.

1.4.3 The Concept of Steady State

The concept of a steady state shows up quite a bit in fluid mechanics. By **steady state**, we mean that the values of the flow variables in the control volume do not change in time. In the example of the continuity equation, a steady state situation implies one where the density does not change with time. Note, that it could still change with space.

Mathematically, let's consider the continuity equation in integral form at steady state:

$$\iiint_V \overset{0}{\cancel{\frac{\partial \rho}{\partial t}}} dV + \oiint_A \rho \vec{V} \cdot \vec{n} dA = 0$$
$$\therefore \oiint_A \rho \vec{V} \cdot \vec{n} dA = 0$$

where the $\therefore$ means therefore. If we assume that the density (ρ) and the velocity ($\vec{V}$) do not vary within a cross-sectional area, then we can utilize this equation for a lot of practical type of situations. For instance, suppose we take the example given in Figure 1.9 of flow in a converging nozzle. If we define our control

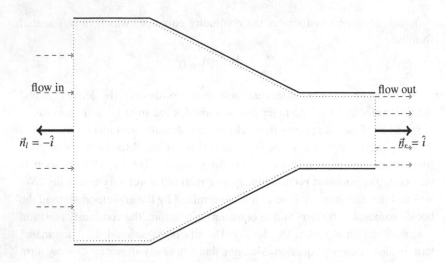

Figure 1.9　Flow in a nozzle.

volume to be inside the dotted line of Figure 1.9, and we define the flow coming in from the left as "*l*" and going out from the right as "*r*," we have:

$$\oiint_A \rho \vec{V} \cdot \vec{n} dA = 0$$

$$\implies \rho_l \vec{V}_l \cdot \vec{n}_l A_l + \rho_r \vec{V}_r \cdot \vec{n}_r A_r = 0$$

(1.37)

where $\implies$ means "implies" the following. The subscripts "*l*" and "*r*" denote values at the left and right faces, respectively. Notice that we were able to easily integrate because we assumed that density and velocity do not depend on A. Also note that we are ignoring the top and bottom surfaces because there is no flow coming in or going out of those surfaces.

Breaking out the $\vec{V} \cdot \vec{n}$ terms in Equation 1.37 leads to:

$$
\begin{aligned}
\rho_l \vec{V}_l \cdot \vec{n}_l A_l + \rho_r \vec{V}_r \cdot \vec{n}_r A_r &= \rho_l \left(u_l \hat{i} + v_l \hat{j} + w_l \hat{k} \right) \cdot \left(n_{lx}\hat{i} + n_{ly}\hat{j} + n_{lz}\hat{k} \right) A_l \\
&\quad + \rho_r \left(u_r \hat{i} + v_r \hat{j} + w_r \hat{k} \right) \cdot \left(n_{rx}\hat{i} + n_{ry}\hat{j} + n_{rz}\hat{k} \right) A_r \\
&= \rho_l \left(u_l \hat{i} + v_l \hat{j} + w_l \hat{k} \right) \cdot \left(-\hat{i} + 0\hat{j} + 0\hat{k} \right) A_l \\
&\quad + \rho_r \left(u_r \hat{i} + v_r \hat{j} + w_r \hat{k} \right) \cdot \left(\hat{i} + 0\hat{j} + 0\hat{k} \right) A_r \\
&= -\rho_l u_l A_l + \rho_r u_r A_r.
\end{aligned}
$$

(1.38)

Equating Equation 1.38 and Equation 1.37, it appears that the value of $\rho u A$ going into the nozzle is equal to the $\rho u A$ going out of the nozzle. In other words, for the flow in our nozzle, we have the simple relationship:

$$\rho_l u_l A_l = \rho_r u_r A_r. \tag{1.39}$$

This relationship applies when the flow is at a steady state and when the velocity and density do not vary within the cross-sectional areas. We can do a simple example with this equation to get a feel for how it works.

Example Suppose water is passing through a nozzle with a circular cross-section and a velocity of u_l. What is the relationship for the velocity of water at the exit (u_r) if the exit radius is half that of the entrance radius, assuming the density of the water does not change as it passes through the system?

Solution The solution to this problem is straightforward. Using the geometry given in Figure 1.9, we can solve Equation 1.39 for u_r to get:

$$u_r = \frac{\rho_l u_l A_l}{\rho_r A_r}.$$

The ρ_l and ρ_r cancel each other out since they are equal in this case (constant density). If we define the radius of the exit of the converging nozzle to be R_r and the radius of the entrance to be $R_l = 2R_r$, we get:

$$u_r = u_l \frac{A_l}{A_r}$$

$$u_r = u_l \frac{\pi R_l^2}{\pi R_r^2}$$

$$u_r = u_l \frac{\pi (2R_r)^2}{\pi R_r^2}$$

$$\therefore u_r = 4u_l$$

Thus the exit velocity (or the velocity at the right edge of the converging nozzle) would be four times that of the entrance velocity.

Of course, we can have slightly more complicated situations where the density of the flow also changes as it passes through the nozzle. This might be more likely to occur if the fluid passing through the nozzle is a gas, such as air. For problems of this type, where there are density changes in the flow, typically a thermodynamic equation of state is needed. A thermodynamic **equation of state** is an equation that provides a relationship between different thermodynamic variables (such as pressure, temperature, and density).[9] A common equation of state for air is the ideal gas equation of state given by:

$$p = \rho R T \tag{1.40}$$

---

[9] Later in the text, we will discuss situations where density and pressure are not to be considered thermodynamic variables. This will be the case in what is known as incompressible flow.

where p is the pressure of the gas, T is the temperature of the gas (in absolute units), and R is a gas constant for the particular type of gas. The unit of pressure is typically pascal, Pa, and is a force per area unit (i.e., $Pa = N/m^2$, where N is a newton and m is meter)). The unit of temperature must be in absolute temperature, e.g., kelvin (K). The gas constant is related to the universal gas constant, $\bar{R}$, via:

$$R = \frac{\bar{R}}{M} \qquad (1.41)$$

where M is the molecular weight for the gas. The universal gas constant has a value of:

$$\bar{R} = 8314 \ \frac{J}{kmolK}.$$

The next example uses an equation of state.

Example Suppose you have an ideal gas passing through a nozzle with a circular cross-section and an inlet velocity of 1.0 m/s. The density of the incoming gas is 1.25 kg/m^3. The exit radius of the nozzle is half that of the entrance nozzle. What is the exit velocity of the flow if the exit temperature is 350 K and the exit pressure is 10^5 Pa? You may assume the system is in a steady state with an ideal gas that has a molecular weight of 28 kg/kmol.

Solution The situation is very similar to the example we did with water, except this time we are using a gas and it appears that the density of the gas may change as it passes through the nozzle. Since it is at a steady state, the situation we have is given by Equation 1.39:

$$\rho_l u_l A_l = \rho_r u_r A_r$$

We can solve for the exit velocity (which would be the velocity on the "right," u_r):

$$u_r = \frac{\rho_l u_l A_l}{\rho_r A_r}.$$

Like the previous example, the exit radius is half that of the entrance radius, thus:

$$u_r = \frac{\rho_l u_l \pi (2R_r)^2}{\rho_r \pi R_r^2}$$

$$\therefore u_r = 4\frac{\rho_l u_l}{\rho_r}.$$

Now we need to find the density on the right side (ρ_r). To find that, we are going to need to use the ideal gas equation of state, Equation 1.40, to get:

$$\rho_r = \frac{p_r}{RT_r}$$

$$\rho_r = \frac{p_r}{\frac{\bar{R}}{M}T_r}$$

$$\rho_r = \frac{10^5 \text{ Pa}}{\frac{8314 \text{ J/kmolK}}{28 \text{ kg/kmol}} 350 \text{ K}}$$

$$\therefore \rho_r = 0.96 \text{ kg/m}^3$$

We can now use this density to calculate the exit velocity:

$$u_r = 4\frac{\rho_l u_l}{\rho_r} = 4\frac{\left(1.25 \text{ kg/m}^3\right)(1 \text{ m/s})}{0.96 \text{ kg/m}^3} = 5.2 \text{ m/s}.$$

We have seen two situations of analyzing flow using the basic idea of mass balance. This leads us to our next discussion of incompressible flows.

1.4.4 Incompressible Flow

Density is an interesting property in fluid mechanics. If a constant density is assumed (which is a fairly common assumption), the resulting governing equations will be what are called the incompressible governing equations. We will define more formally what an incompressible flow is as we progress in this book (there will be a number of definitions) but the definition we will use for now is one of a constant density (which is the most restrictive definition). Constant density in this case means the value of density is the same everywhere throughout the flow. For such situations, the density is often obtained via a look-up table for a given fluid at a given temperature. Effectively, this implies that an incompressible flow is very much in line with an incompressible fluid (which more formally implies that the density of a fluid does not change with pressure). However, as we will see in the last chapter, you can have an incompressible flow with a compressible fluid (such as air), provided that the speed of flow is low enough.

Nevertheless, in situations where the density is constant, the continuity equation given in Equation 1.23 (i.e., $\frac{\partial \rho}{\partial t} + \vec{\nabla} \cdot \left(\rho\vec{V}\right) = 0$) can be simplified (since the time derivative goes away and the density "divides" out of the divergence) to get:

$$\boxed{\vec{\nabla} \cdot \vec{V} = 0} \tag{1.42}$$

Equation 1.42 is the resulting continuity equation in differential form for an **incompressible flow**. We will see as we progress that all of the definitions of an incompressible flow lead to Equation 1.42.

Equation 1.42 states that the divergence of the velocity field is zero. Sometimes you might hear that the velocity is divergence-free or solenoidal in an incompressible flow. So, what does a divergence-free velocity field mean? We will discuss a couple different physical interpretations of the divergence-free velocity field in this book. However, at this point, to give our first insight into this equation, consider Equation 1.42 in Cartesian coordinates:

$$\frac{\partial u}{\partial x} + \frac{\partial v}{\partial y} + \frac{\partial w}{\partial z} = 0 \tag{1.43}$$

If we just consider, for simplicity, a one-dimensional flow in the x-direction, then our continuity equation becomes a very simple expression:

$$\frac{\partial u}{\partial x} = 0$$

So, what does this mean? This simply means that the velocity in the x-direction (u) does not change along the x-coordinate. If we approximate this simple derivative algebraically, that is, as a slope (i.e., the change in the velocity divided by the change in x), we get:

$$\frac{u_r - u_l}{\Delta x} = 0$$

$$\therefore u_r = u_l$$

where u_l is the velocity at the left edge of a Cartesian volume and u_r is the velocity at the right edge of a Cartesian volume as shown in Figure 1.10a. This means, effectively, that what goes in on one side must come out on the other side.

We can extend this idea to three dimensions as shown in the Cartesian control volume in Figure 1.10b. We can approximate Equation 1.43 as an algebraic equation instead of a differential equation by replacing the derivatives with slopes via:

$$\frac{u_r - u_l}{\Delta x} + \frac{v_t - v_b}{\Delta y} + \frac{w_f - w_{bk}}{\Delta z} = 0$$

where v_t and v_b are the flow velocities in the y-direction at the top and bottom surface, respectively. In addition, w_f and w_{bk} are the flow velocities in the z-direction at the front and the back surface of the Cartesian volume, respectively. If we assume Δx, Δy, and Δz are equal, then:

$$u_r - u_l + v_t - v_b + w_f - w_{bk} = 0$$

$$\therefore u_l + v_b + w_{bk} = u_r + v_t + w_f$$

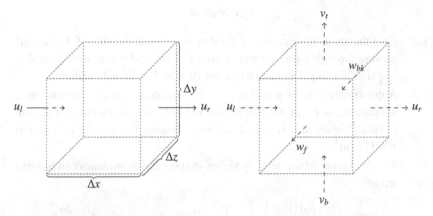

(a) Flow only in the *x*-direction.   (b) Flow in three all three dimensions.

Figure 1.10 Velocity components of the flow passing through a Cartesian volume.

This implies that the total flow coming in is equal to the total flow going out. Another way of saying the same thing is that there are no sources or sinks of mass inside the volume.

Notice that $\vec{\nabla} \cdot \vec{V} = 0$ is not an equation for density (density is not even in the equation), unlike the full continuity equation, $\frac{\partial \rho}{\partial t} + \vec{\nabla} \cdot \left(\rho \vec{V} \right) = 0$. It turns out that Equation 1.42 is really a constraint equation. It ends up constraining the values of pressure in an incompressible flow. Thus, you may sometimes hear Equation 1.42 being referred to as the incompressibility constraint equation.

Situations where the density is dependent on the flow conditions, particularly on the pressure, show up in what are called compressible flows. The continuity equation for compressible flows is given by Equation 1.23, or Equation 1.20 in integral form. The continuity equation in compressible flows is an equation for density and it is explicitly tied to the velocity field. As a result, density is treated like a flow variable (i.e., an unknown whose value is to be obtained by calculating the continuity equation) in compressible flows. In addition, the density, pressure, and temperature of the flow are all related through a thermodynamic equation of state in compressible flows. Thus, the density can also be considered a thermodynamic variable in compressible flows. This is in contrast to incompressible flows, where the density is usually not considered an unknown flow variable but is instead just an input parameter (i.e., you get the value of density from a table in a book).

Problems

1.1   Find the mass flow rate of a fluid at a constant density of 1.2 kg/m^3 passing through a surface whose area is 1 m^2 and whose normal is $\vec{n} = \frac{1}{\sqrt{14}}\left(\hat{i} + 2\hat{j} + 3\hat{k}\right)$. The velocity of the flow is: $\vec{V} = 2\hat{i} + 3\hat{j} + 0\hat{k}$ m/s.

1.2   Approximate how long will it take, in minutes, to fill up a bathtub with dimensions of 1.3 m × 0.6 m × 0.4 m if water is coming out of a 40 mm diameter faucet at a speed of 1.2 m/s. Assume the density of water is 1000 kg/m^3.

1.3   The divergence of the velocity vector in spherical coordinates can be written as:

$$\vec{\nabla} \cdot \vec{V} = \frac{1}{r^2}\frac{\partial\left(r^2 V_r\right)}{\partial r} + \frac{1}{r\sin(\theta)}\frac{\partial}{\partial\theta}\left(V_\theta \sin(\theta)\right) + \frac{1}{r\sin(\theta)}\frac{\partial V_\phi}{\partial\phi}$$

where V_r, V_θ, and V_ϕ are the velocity coordinates in the r-, θ-, and ϕ-directions, respectively. Determine if a flow with the following flow field velocity is incompressible:

$$V_r = -U\cos(\theta)\left(1 - \frac{3R}{2r} + \frac{R^3}{2r^3}\right)$$

$$V_\theta = U\sin(\theta)\left(1 - \frac{3R}{4r} - \frac{R^3}{4r^3}\right)$$

$$V_\phi = 0$$

where R and U are constants (note, R is not the gas constant in this problem).

1.4   An incompressible fluid with density $\rho = 1000$ kg/m^3 travels through a channel with a rectangular cross-section of dimensions 25 mm by 30 mm. The average velocity of the flow in this portion of the channel is 1 m/s. If the flow is at a steady state, what will the velocity be if the rectangular cross-section increases to 50 mm by 50 mm? **Start from the integral form of the continuity equation.**

1.5   Air in a pipe with a diameter of 10 cm starts out at a temperature of 700 K and pressure of 4×10^5 Pa. The initial flow velocity is 10 meters per second. If the pipe diameter contracts to 5 cm with the air speeding up to 115 m/s and the temperature decreases to 500 K, what is the pressure after the contraction? You can assume the molecular weight of air is 28.97 kg/kmol and a steady state. **Start from the integral form of the continuity equation.**

1.6 The velocity profile of flow in a pipe in the z-direction (V_z) is given by:

$$V_z = \frac{\Delta p R^2}{4\mu L}\left(1 - \frac{r^2}{R^2}\right)$$

where r is the radial component (in cylindrical coordinates), R in this case is the radius of the pipe, L is the length of the pipe, Δp is considered to be a pressure difference between the entrance and exit of the pipe, and μ is something called the dynamic viscosity. Obtain an expression for the mass flow rate through the pipe.

1.7 If the velocity profile of flow in a channel is given by: $u = \frac{U_\infty}{H}y$, what is the mass flow rate through the channel per length into the page if the height of the channel is H?

1.8 Given the following velocity vector: $\vec{V} = Cy\cos(5x)\hat{i} + D\sin(5x)y^2\hat{j}$, what do C and D need to be in order for this vector field to be considered incompressible flow?

1.9 The flux of a quantity given by the vector, $\vec{f} = 5xy\hat{i} + 10y\hat{j} + 0\hat{k}$, passes through a cube with dimensions (in x-, y-, and z-directions) of $2 \times 1 \times 1$ (assume the cube "starts" at the origin). Find the value of the area integral, $\oiint_A \vec{f} \cdot \vec{n}dA$.

1.10 Sketch the velocity vector given by: $\vec{V} = -\sin(y)\hat{i} + \sin(x)\hat{j}$.

2

The Material Derivative:
The First Step to the Navier–Stokes Equations

In the last chapter, we developed the continuity equation by applying mass conservation to a moving fluid. Now we would like to think about how we would apply a force balance ($\vec{F} = m\vec{a}$) to a moving fluid, which will lead to the Navier–Stokes equations. Before we perform such a force balance, let us think for a moment about what we did in the last chapter. The continuity equation we developed in the last chapter was the result of a mass balance on control volume fixed in space. That is to say, as an observer of the flow, we fixed our eyes on a region of space and watched the fluid moving past us. We kept track of the rate of mass in and the rate of mass out of our fixed control volume and calculated how much mass was accumulating or leaving the volume in a given period of time. This approach worked well and was somewhat intuitive for mass conservation. However, what about using the method developed in Chapter 1 for a force balance? You most certainly can do so, as we will see Chapter 4, but it might not be quite as obvious how to perform a force balance on a fixed volume with a "flux" of fluid coming in and going out. After all, a more standard approach for a force balance is to calculate the net force acting on a moving object of fixed mass. As a result, the more intuitive approach for performing a force balance on our fluid would probably be doing something along the lines of calculating a net force on a "particle" of fluid. This "viewpoint" is different than what we did with mass balance in the last chapter. These are two different ways of looking at fluid motion and you can study the conservation principles using either of these approaches. These two approaches have a name, one is the Lagrangian description (which is following a fluid particle around) and the other is the Eulerian description (watching fluid pass through a fixed control volume). Understanding each of these descriptions (or frameworks) of flow is needed to fully understand the dynamics of fluids.

This chapter is devoted to diving into the differences between the two descriptions of fluid motion. Understanding this chapter will help us tremendously in our understanding of the following chapters when we discuss the Navier–Stokes and energy equations. This chapter will introduce and discuss a new type of derivative, the material derivative. It is extremely important to understand this derivative before we tackle the Navier–Stokes equations themselves.

2.1 Lagrangrian and Eulerian Descriptions

The Lagrangian description describes the motion of the flow by assuming an observer is located on a parcel of fluid of fixed mass as it travels through space. The Lagrangian description is typically what a student is first introduced to when studying particle motion and dynamics using Newtonian mechanics. The "particle" in the case of the Lagrangian description of fluid mechanics is called a fluid element (or sometimes a material element). As the fluid element travels through space, whatever constituents that make it up (e.g., atoms or molecules) must remain the same. Bear in mind that it is still assumed that the number of atoms contained inside a fluid element is large enough so that the fluid element can still be considered a continuum (i.e., individual atomic effects can be ignored).

The second description of fluid flow is the Eulerian description. In the Eulerian description, an observer is positioned at a point in space and "watches" the fluid flow as it passes them at that fixed point. This is what we did in the last chapter with mass balance and the development of the continuity equation. In the Eulerian description, the concept of fluxes (e.g., mass flux, momentum flux, energy flux) through a surface is often utilized. In addition, the Eulerian description utilizes a control volume as the system of interest.

Figure 2.1 illustrates the difference between the two descriptions of fluid flow by again looking at flow in a nozzle. The nozzle diameter reduces as the flow goes downstream, thus accelerating (under typical circumstances) the flow. The Lagrangian description is depicted in Figure 2.1 by a little stick figure person sitting on top of a sphere (a fluid element). The stick figure person stays with that fluid element as it travels through the nozzle. The fluid element (and hence the stick figure person) will experience an acceleration as it moves through the nozzle. Note that even though the fluid element shown in Figure 2.1 is depicted as a sphere, the fluid element does not have to be a sphere. It may be any shape and may deform or change its velocity (or temperature, pressure, density, etc.) as it moves from point to point. At this point in our discussion, the fluid element will be considered small enough so that the properties within the fluid element remain uniform. That is to say, the temperature, pressure, etc. have a well-defined value within the fluid element.

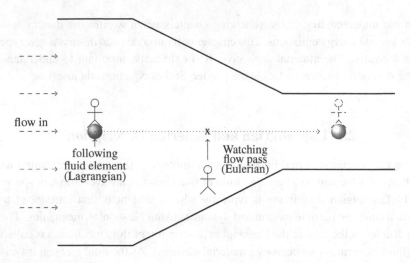

Figure 2.1 Illustration of Lagrangian and Eulerian descriptions.

The Eulerian description is depicted by the stick person in the middle of the nozzle looking at flow passing a point (marked by 'x') in Figure 2.1. The stick person would be able to measure various properties (temperature, pressure, velocity, etc.) at the point 'x'.

Both Lagrangian and Eulerian descriptions can be used when deriving the governing equations. The Langrangian descriptions and the Eulerian descriptions can be related to one another through what is known as the material derivative. There are many approaches to illustrating what a material derivative is but probably the best approach to use in order to understand the physical meaning is through the use of a Taylor series. The approach followed here is similar to an approach found in Dr. John Anderson's wonderful book on computational fluid mechanics.[1]

To start off, suppose we are tracking the temperature of a small fluid element, such as the one shown in Figure 2.2, as it moves through the nozzle horizontally in the x-direction from point 1 (x_1) at time, t_1, to point 2 (x_2) at some later time, t_2. The value of the temperature at (x_2, t_2), given by T_2, can be written in terms of the value of the temperature at (x_1, t_1), given by T_1, using the Taylor series via:

$$T_2 = T_1 + \left(\frac{\partial T}{\partial t}\right)_1 \Delta t + \left(\frac{\partial T}{\partial x}\right)_1 \Delta x + \frac{1}{2}\left(\frac{\partial^2 T}{\partial t^2}\right)_1 \Delta t^2 + \frac{1}{2}\left(\frac{\partial^2 T}{\partial x^2}\right)_1 \Delta x^2 + ... \quad (2.1)$$

---

[1] The reference for this book can be found in the Further Reading section.

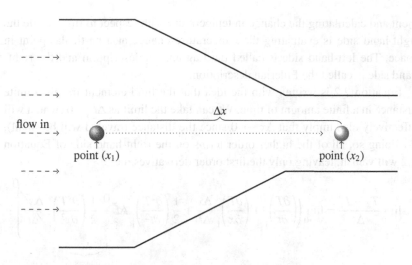

Figure 2.2 Tracking a fluid element as it moves from point (x_1) to point (x_2) in a time $\Delta t = t_2 - t_1$.

The Δx is the distance from point 1 to point 2, i.e., $\Delta x = x_2 - x_1$. Similarly, $\Delta t = t_2 - t_1$ is the time it took to get from point 1 to point 2. As a reminder, recall from calculus that the Taylor series is a way to determine the value of a function at a particular point, given the value of that function at another point and how that function changes (which is calculated through the derivatives) at that other point. So, if we are considering our current scenario, we can calculate the temperature at point 2 (T_2) by knowing the temperature at point 1 (T_1) and by knowing how the temperature varies (via the derivatives) at point 1. You will notice that the derivatives in the Taylor series given above are evaluated at point 1 (the subscript 1 in the derivatives of the Taylor series indicates that the values of the derivatives are evaluated at point 1). Rearranging Equation 2.1 by subtracting T_1 from both sides and dividing by Δt gives us:

$$\frac{T_2 - T_1}{\Delta t} = \left(\frac{\partial T}{\partial t}\right)_1 + \left(\frac{\partial T}{\partial x}\right)_1 \frac{\Delta x}{\Delta t} + \frac{1}{2}\left(\frac{\partial^2 T}{\partial t^2}\right)_1 \Delta t + \frac{1}{2}\left(\frac{\partial^2 T}{\partial x^2}\right)_1 \frac{\Delta x^2}{\Delta t} + \dots \quad (2.2)$$

Notice that the left-hand side of Equation 2.2 is calculating the temperature difference of the fluid element divided by the time difference as it moves from point 1 to 2. In other words, it is following the fluid element from point 1 to point 2 and "tracking" the temperature difference (with respect to time) of that fluid element. Contrast that with what is on the right hand side of Equation 2.2. The values of the derivatives on the right-hand side are evaluated only at point 1. You can think of the right-hand side being calculated as though an observer is stationary at point 1. To reiterate, the left-hand side is following a fluid element and calculating the change in temperature with respect to time, while the right-hand side is evaluating the temperature change at a particular point in space. The left-hand side is called the Lagrangian description and the right-hand side is called the Eulerian description.

Equation 2.2 is written with the idea that the fluid element travels a finite distance in a finite amount of time. We can take the limit as $\Delta t \to 0$ (which will effectively also imply that $\Delta x \to 0$ since the distance traveled will be small). By doing so, all of the higher order terms on the right-hand side of Equation 2.2 will vanish, leaving only the first order derivatives, i.e.:

$$\lim_{\Delta t \to 0} \frac{T_2 - T_1}{\Delta t} = \lim_{\Delta t \to 0} \left(\left(\frac{\partial T}{\partial t} \right)_1 + \left(\frac{\partial T}{\partial x} \right)_1 \frac{\Delta x}{\Delta t} + \frac{1}{2} \left(\frac{\partial^2 T}{\partial t^2} \right)_1 \cancel{\Delta t}^{\,0} + \frac{1}{2} \left(\frac{\partial^2 T}{\partial x^2} \right)_1 \cancel{\frac{\Delta x^2}{\Delta t}}^{\,0} \right.$$

$$+ \underbrace{....}_{\substack{\text{goes} \\ \text{to zero}}}$$

$$(2.3)$$

Notice that the limit of $\frac{\Delta x}{\Delta t}$ becomes the velocity in the x-direction (u). In addition, the right-hand side no longer needs to distinguish that the derivatives be evaluated at point 1 since the limit implies that point 1 approaches point 2. This leads us to the following equation:

$$\lim_{\Delta t \to 0} \frac{T_2 - T_1}{\Delta t} = \frac{\partial T}{\partial t} + \frac{\partial T}{\partial x} u \qquad (2.4)$$

We are still not quite done yet, but we are getting there. We still kept the limit on the left-hand side. Instead of keeping the limit, we are going to replace it with a derivative. However, instead of replacing the left-hand side with an ordinary derivative, we are going to use a standard fluid dynamics convention and replace it with a derivative that uses an uppercase 'D' (instead of the usual lowercase 'd'). By doing this, our equation becomes:

$$\frac{DT}{Dt} = \frac{\partial T}{\partial t} + \frac{\partial T}{\partial x} u \quad \text{where} \quad \left(\frac{DT}{Dt} = \lim_{\Delta t \to 0} \frac{T_2 - T_1}{\Delta t} \right) \qquad (2.5)$$

Moving the velocity u to the other side of the partial derivatives gives us the standard way of writing this expression (in one dimension), i.e.:

$$\frac{DT}{Dt} = \frac{\partial T}{\partial t} + u \frac{\partial T}{\partial x}. \qquad (2.6)$$

Equation 2.6 is assuming the flow is only in the x-direction. We can do the exact same Taylor series expansion procedure for a fluid element that moves in three dimensions from an initial position of (x_1, y_1, z_1) at time t_1 to a final position of (x_2, y_2, z_2) at time t_2. In doing so, we get analogous terms for the y-direction and z-direction as we did for the x-direction. Thus, we end up with two new terms: the velocity in the y-direction (i.e., v) multiplied by the derivative of T with respect to y, and a velocity in the z-direction (i.e., w) multiplied by the derivative of T with respect to z. Therefore, the three dimensional version of Equation 2.6 is:

$$\frac{DT}{Dt} = \frac{\partial T}{\partial t} + u\frac{\partial T}{\partial x} + v\frac{\partial T}{\partial y} + w\frac{\partial T}{\partial z}. \tag{2.7}$$

We can also write this equation in coordinate-free notation using the nabla symbol:

$$\frac{DT}{Dt} = \frac{\partial T}{\partial t} + \vec{V} \cdot \vec{\nabla} T. \tag{2.8}$$

Equation 2.8 provides us with a relationship for the material derivative. In this case, the material derivative is that of temperature. The left-hand side of Equation 2.8 is the material derivative in "Lagrangian form," while the right-hand side is the "Eulerian form." We can write the material derivative as an "operator," with "( )" being where a variable should go:

$$\boxed{\frac{D()}{Dt} = \frac{\partial()}{\partial t} + \vec{V} \cdot \vec{\nabla}()} \tag{2.9}$$

Much like a normal derivative, Equation 2.9 has to have some variable to "operate" on. In fluid mechanics, the material derivative can operate on any field variable (e.g., density, temperature, velocity, pressure).

The terms of the material derivative have common names. The names of the terms, along with the Lagrangian and Eulerian descriptions, are labeled below:

$$\underbrace{\frac{D()}{Dt}}_{\substack{\text{Lagrangian}\\\text{description}}} = \underbrace{\underbrace{\frac{\partial()}{\partial t}}_{\substack{\text{local time}\\\text{derivative}}} + \underbrace{\vec{V} \cdot \vec{\nabla}()}_{\substack{\text{advective}\\\text{transport}}}}_{\substack{\text{Eulerian}\\\text{description}}}$$

The left-hand side of the equal sign is the **material derivative in a Lagrangian description** and the right-hand side of the equal sign is the **material derivative in an Eulerian description**. The first term on the right is typically called a **local time derivative**. Local meaning that the derivative is taken at a fixed point in space. The second term on the right goes by several names, such as the advective term, the convective term, the convective transport term, the advective transport term, the convective acceleration term, the convective derivative, among others. In this text, however, we will generally refer to it as either the **advective term** or the **advective transport term**.[2]

The advective term can be broken out in Cartesian coordinates much like a simple dot product, i.e.:

$$\vec{V} \cdot \vec{\nabla} = \left(u\hat{i} + v\hat{j} + w\hat{k} \right) \cdot \left(\hat{i} \frac{\partial}{\partial x} + \hat{j} \frac{\partial}{\partial y} + \hat{k} \frac{\partial}{\partial z} \right)$$
$$= u \frac{\partial}{\partial x} + v \frac{\partial}{\partial y} + w \frac{\partial}{\partial z}. \tag{2.10}$$

Thus, in Cartesian coordinates, the material derivative using the Eulerian description (i.e., right-hand side) would be:

$$\frac{D}{Dt} = \frac{\partial ()}{\partial t} + u \frac{\partial ()}{\partial x} + v \frac{\partial ()}{\partial y} + w \frac{\partial ()}{\partial z}. \tag{2.11}$$

Notice that the material derivative operation does not contain any base vectors $(\hat{i}, \hat{j}, \hat{k})$. Thus applying the material derivative to a scalar (that is, a quantity that just has a magnitude such as density) will result in a scalar and operating on a vector, like the velocity vector, will result in a vector. If we were to apply the material derivative to the velocity vector $(\vec{V})$ we would get:

$$\frac{D\vec{V}}{Dt} = \frac{\partial \vec{V}}{\partial t} + \vec{V} \cdot \vec{\nabla}\vec{V}. \tag{2.12}$$

In Cartesian coordinates, the material derivative of the velocity vector becomes:

$$\frac{D}{Dt} \underbrace{\begin{pmatrix} u \\ v \\ w \end{pmatrix}}_{\vec{V}} = \overbrace{\frac{\partial}{\partial t} + \underbrace{u \frac{\partial}{\partial x} + v \frac{\partial}{\partial y} + w \frac{\partial}{\partial z}}_{\vec{V} \cdot \vec{\nabla}}}^{\text{operates on each velocity component}} \underbrace{\begin{pmatrix} u \\ v \\ w \end{pmatrix}}_{\vec{V}}$$

$$\rightarrow \begin{pmatrix} \frac{Du}{Dt} \\ \frac{Dv}{Dt} \\ \frac{Dw}{Dt} \end{pmatrix} = \begin{pmatrix} \frac{\partial u}{\partial t} + u \frac{\partial u}{\partial x} + v \frac{\partial u}{\partial y} + w \frac{\partial u}{\partial z} \\ \frac{\partial v}{\partial t} + u \frac{\partial v}{\partial x} + v \frac{\partial v}{\partial y} + w \frac{\partial v}{\partial z} \\ \frac{\partial w}{\partial t} + u \frac{\partial w}{\partial x} + v \frac{\partial w}{\partial y} + w \frac{\partial w}{\partial z} \end{pmatrix} \tag{2.13}$$

The material derivative of velocity will show up in the Navier–Stokes equations and is nothing more than an equation for the acceleration of the flow. The left-hand side of Equation 2.12 is essentially the same as the derivative

---

[2] Incidentally, some call the material derivative the convective derivative. There are other names for the material derivative as well. Other common names are: substantial derivative, total derivative, and the hydrodynamic derivative.

you would see in Newtonian mechanics when studying moving point particles. The reason for the uppercase 'D' instead of a lowercase 'd' is, in some ways, nothing more than a common convention. Some books still use a lowercase 'd' when writing the Lagrangian form of the material derivative. However, most use the uppercase 'D' to distinguish the material derivative from the ordinary derivative. The ordinary derivative is a derivative where the dependent variable is only a function of one independent variable. The material derivative, on the other hand, is specifically used to describe the change of a flow variable with respect to time of a fluid element as it moves through space.

To get an idea of the contrast between the material derivative in the two different descriptions (Lagrangian and Eulerian), let's revisit the situation where we have flow in a converging nozzle as was shown in Figure 2.1. The flow is coming in from the left and moving to the right through the nozzle. We probably know, from experience, that by reducing the area through which the flow travels, the velocity of the flow will increase. Let's take a look at the two different approaches (Lagrangian and Eulerian) for this scenario.

First, consider the Lagrangian approach where our little stick figure is located on the fluid element and tracks (or measures) the velocity of the element as it moves. As you would expect, the fluid element will accelerate as it moves through the nozzle. Therefore, on average, we can expect that the material derivative of velocity (which is just the acceleration of the fluid element) will be greater than 0, i.e., $\frac{D\vec{V}}{Dt} > 0$, as it makes its way through the nozzle.

Now consider the Eulerian description. This time, the stick figure will just sit in the middle of the nozzle and watch the fluid move past a point. We can have the stick figure take some velocity measurements at various time intervals. The change in the velocity measurements at the various time intervals provides the local time derivative. Thus if the velocity values that are measured do not change with time (that is, if the velocity of the flow at 1 minute is the same as the velocity at 10 minutes, which is the same as the velocity 10 hours, etc.), then the local time derivative will be zero. In other words:

$$\frac{D\vec{V}}{Dt} = \overset{0}{\cancel{\frac{\partial \vec{V}}{\partial t}}} + \vec{V} \cdot \vec{\nabla}\vec{V}. \tag{2.14}$$

This implies that the acceleration of the fluid that occurs as it flows through the nozzle does not come from a time derivative of velocity in the Eulerian description (which is normally what one would expect for acceleration). Instead, the acceleration comes from the advective term, which contains spatial (i.e., x, y, z) derivatives of velocity. This would mean that in order for our little stick figure sitting in the middle of the nozzle to measure the acceleration of the flow, they would need additional velocity measurements surrounding the point 'x' in order to get velocity values at different spatial locations. This will enable them to calculate the spatial derivatives in the advective transport term.

Of course, there could be a situation in which the velocity measurements also change in time (leaving $\frac{\partial \vec{v}}{\partial t} \neq 0$), in which case both the local time derivative and the advective term are not equal to zero. As mentioned in the last chapter, the special case when the local time derivative can be assumed to be zero is called a **steady state** condition.

We are now going to further investigate the material derivative by looking at two relatively simple partial differential equations, the advection equation and the inviscid Burgers' equation.

2.2 The Advection and Inviscid Burgers' Equation

It will be useful to look at the simplest equation involving a material derivative. That is, one where the material derivative is set to zero. If the material derivative of velocity equals zero, the resulting equation is called the Burgers' equation. More specifically, the inviscid Burgers' equation. If the material derivative of any other property (e.g., temperature, pressure, density, etc.) is zero, the resulting equation is called an advection equation. The advection equation is easier to follow and understand (as we shall see), so that is where we will begin.

2.2.1 Advection Equation

Advection, by definition, means motion or transport of some quantity or property by means of a moving fluid. Our property of choice to study the advection equation will be temperature since advection of temperature (or more precisely, energy) will show up in the energy equation.

Setting the material derivative of temperature to zero leads to the following equation in the Eulerian description:

$$\frac{\partial T}{\partial t} + \vec{V} \cdot \vec{\nabla} T = 0. \tag{2.15}$$

Working in Cartesian coordinates and assuming that only the x-velocity (i.e., u) is non-zero leads us to a one-dimensional form of the advection equation:

$$\frac{\partial T}{\partial t} + u\frac{\partial T}{\partial x} = 0. \tag{2.16}$$

This is a partial differential equation (PDE) because the dependent variable, T, has derivatives with respect to more than one independent variable. In this case, the derivatives of T are with respect to x and t. In general, solving differential equations usually involves a guess for a solution. The guessed solution is then "plugged" into the differential equation to see if it is, in fact, a solution that works. Of course, in your differential equations class, there are methods that you have been taught that enable you to solve a specialized set of differential equations. However, most of those methods originated with a guessed solution by a mathematician or physicist at some point in history. If a guess for a solution is deemed successful, then a method can potentially be "backed" out. Unfortunately, even though they show up everywhere in physics and engineering, most partial differential equations do not have general solutions because of their complexity and must be either simplified greatly or solved on a computer using numerical techniques. The advection equation, however, is one of the simpler PDEs and does have a general solution.

The solution to the 1D advection equation for T (or more formally, $T(x, t)$) is:

$$T(x, t) = T(x - ut, 0). \tag{2.17}$$

You can verify that Equation 2.17 is a solution to the 1D advection equation by plugging it into the 1D advection equation, Equation 2.16. Plugging $T(x - ut, 0)$ for T in the advective term $\left(u\dfrac{\partial T}{\partial x}\right)$ gives:

$$u\frac{\partial T}{\partial x} = u\frac{\partial T(x - ut, 0)}{\partial x} \tag{2.18}$$

That was easy enough since it was just a simple substitution (e.g., $T(x - ut, 0)$ for T).

The local time derivative is a little more tricky. Plugging the solution, that is, Equation 2.17, into the local time derivative gives:

$$\frac{\partial T}{\partial t} = \frac{\partial T(x - ut, 0)}{\partial t} = \underbrace{\frac{\partial T(\overbrace{x - ut}^{\hat{x}}, 0)}{\partial \hat{x}} \underbrace{\frac{\partial \hat{x}}{\partial t}}_{-u}}_{\text{chain rule}} = \underbrace{-u\frac{\partial T(x - ut, 0)}{\partial x}}_{\text{result}} \tag{2.19}$$

At first glance it might appear that taking the time derivative of the solution with respect to time, t, might lead to zero because the time parameter in the solution (i.e., $T(x - ut, 0)$) is zero. However, the other parameter in the solution is $x - ut$, which is a function of time. Thus, the chain rule from calculus was used with $\hat{x}$ set equal to $x - ut$. As you can see, the proposed solution ($T = T(x - ut, 0)$) is a solution to the advection equation since the local time derivative and the advective term are negatives of each other (i.e., $\pm u\frac{\partial T(x-ut,0)}{\partial x}$) and cancel each other out.

The solution to the advection equation relies on the initial condition of the system. The **initial condition** indicates what the values of the system property

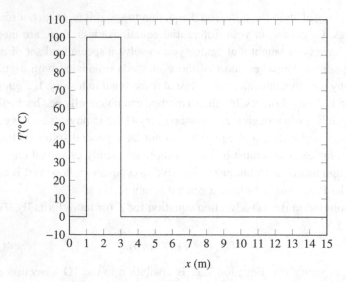

Figure 2.3 Initial temperature distribution for the advection example.

(in this case, temperature) are at some initial time (usually a time, t, of 0). Mathematically, the initial condition might be written: at $t = 0$, $T = T(x, 0)$, where $T(x, 0)$ is some initial temperature distribution of the system.

The following example illustrates how the solution to the advection equation works.

Example Suppose we have flow in a channel or pipe.[3] The flow has an initial (at $t = 0$) temperature distribution (T vs x) given by the plot in Figure 2.3. The velocity of the flow is a constant and is solely in the x-direction with $u = 1$ m/s. What happens to the temperature distribution via advection of the flow after 3 seconds, 6 seconds, and 9 seconds?

Solution The equation we are going to be working with is the one-dimensional (in the x-direction) advection equation for temperature:

$$\frac{\partial T}{\partial t} + u\frac{\partial T}{\partial x} = 0$$

As mentioned the solution to this equation is the following: $T = T(x - ut, 0)$. To understand how to apply that equation, let's consider an example by picking a particular "x" point, say at $x = 5.0$ m. If you wanted to calculate

---

[3] Not all fluid dynamics examples deal with the flow in a pipe or channel but their familiarity often makes for useful examples.

Table 2.1 *Table of T values for the advection equation example*

x (m)	T (celsius) (@t = 0) (T=T(x,0))	T (celsius) (@t = 3 s) (T=T(x-u*3,0))	T (celsius) (@t = 6 s) (T=T(x-u*6,0))	T (celsius) (@t = 9 s) (T=T(x-u*9,0))
0	0	0	0	0
1.0	100	0	0	0
2.0	100	0	0	0
3.0	100	0	0	0
4.0	0	100	0	0
5.0	0	100	0	0
6.0	0	100	0	0
7.0	0	0	100	0
8.0	0	0	100	0
9.0	0	0	100	0
10.	0	0	0	100
11.	0	0	0	100
12.	·0	0	0	100

the temperature at $x = 5.0$ m at $t = 3$ s, then $T = T(x - ut, 0) = T(5.0 - 1.0 * 3, 0) = T(2, 0) = 100°$C. You can do the same procedure with multiple x and t points to get an idea of what happens to the temperature distribution as time passes. Table 2.1 provides sample points (note, all values of the initial temperature not shown in Figure 2.3 are assumed to be zero). Figure 2.4 plots the temperature distribution at various times corresponding to Table 2.1. As you probably already suspected, the temperature distribution just translates (or advects) "downstream" to the right, and does so with a velocity of $u = 1$ m/s.

The above approach solved the advection equation for temperature, which is just the material derivative of temperature set to zero, in Eulerian form. That is to say, we calculated the temperature at particular positions in x at different times. Remember, though, that we could have approached this problem using a Lagrangian description. In the Lagrangian description, fluid elements are followed and their properties are evaluated as the fluid element moves. Let's do the same example, except now use a Lagrangian approach. To do this approach, let's mark "fluid elements" on the initial curve, as seen in Figure 2.5. As before, the velocity of the system is a constant in the x-direction of $u = 1$ m/s. Thus, all of the fluid elements will move to the right with a velocity of 1 m/s. What about the temperature of each of the fluid elements? Since the material derivative in the Lagrangian description is just $\frac{DT}{Dt} = 0$, this implies that the temperature of each fluid element does not change as it moves (since the derivative is zero). This will result in the temperature distribution curve staying the same

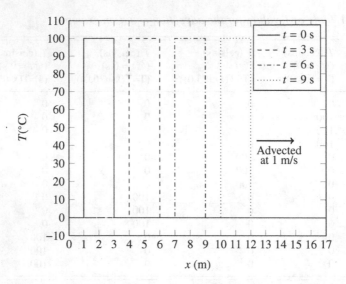

Figure 2.4 Temperature distribution in the advection equation example at various times.

shape, just advecting to the right, as was seen by studying the advection equation via the Eulerian description. Thus instead of calculating the temperature at given x values as we did using the Eulerian form of the advection equation, the Lagrangian form simply says that each point on the initial curve is followed as it moves with a velocity of u. The distance each fluid element moves is just the velocity multiplied by the time elapsed.

We can also plot the trajectories of the fluid elements on a t vs. x curve. The resulting curves obtained are called characteristic curves. The x position is calculated much like one would calculate the position of a point particle, i.e.:

$$x = x_0 + ut$$

where x_0 is the initial position of a fluid element. The characteristic curves are usually drawn with time on the vertical axis, thus rearranging gives t as a function of x:

$$t = \frac{1}{u}(x - x_0)$$

The characteristic curves for the three fluid points marked in Figure 2.5 create lines with a slope of $\frac{1}{u}$, as seen in Figure 2.6. We could have marked any

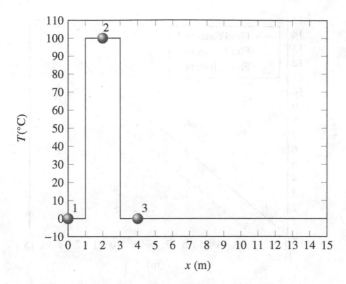

Figure 2.5 Lagrangian approach to the advection example. Three fluid elements are marked as spheres. They simply will move to the right with a velocity of 1 m/s, keeping the same temperature.

number of fluid elements and, at least in the present situation, they would all have the same slope (just different starting positions). Each fluid element travels along its own characteristic curve and the value of the flow property (in this case, temperature) does not change along the curve. Thus, the temperature along the characteristic curve for fluid element 1 and fluid element 3 will be 0°C, and the temperature along the characteristic curve for fluid element 2 will be 100°C.

Characteristic curves are useful tools when solving certain partial differential equations via a method called the method of characteristics. This method is used quite often for partial differential equations that are called hyperbolic. Besides the advection equation, other hyperbolic differential equations are the wave equation and the inviscid Burgers' equation. This book is not going to cover the method of characteristics. However, characteristic curves will still be an invaluable tool when we cover the inviscid Burgers' equation. In the example of the advection equation with a constant velocity, the characteristic curves are just a series of parallel lines. With the inviscid Burgers' equation, you will see that the lines will not be parallel. This will have implications in determining whether a shock wave (a region of discontinuity) will be formed.

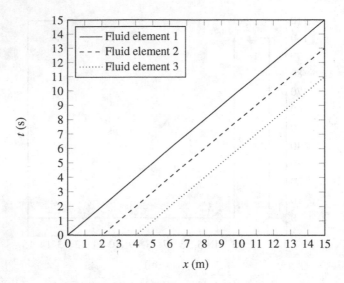

Figure 2.6 Characteristic curves for the advection equation with constant velocity of 1 m/s.

2.2.2 Inviscid Burgers' Equation

We are now in a position to introduce the Burgers' equation. The Burgers' equation comes in two basic forms: an inviscid form and a viscous form. In this chapter, we will look at the inviscid form. We will start by setting the material derivative of velocity to zero, that is:

$$\frac{\partial \vec{V}}{\partial t} + \vec{V} \cdot \vec{\nabla} \vec{V} = 0. \tag{2.20}$$

In one dimension (the x-direction), we obtain the inviscid Burgers' equation:

$$\frac{\partial u}{\partial t} + u \frac{\partial u}{\partial x} = 0. \tag{2.21}$$

The solution for u in the Burgers' equation[4] looks much like the solution for T in the advection equation, that is:

$$u = u(x - ut, 0). \tag{2.22}$$

However, as you can see, there is an issue in that the solution for velocity now depends on the velocity itself. Such a solution is called an implicit

---

[4] At this point, for this chapter, we will use inviscid Burgers' equation and Burgers' equation interchangeably.

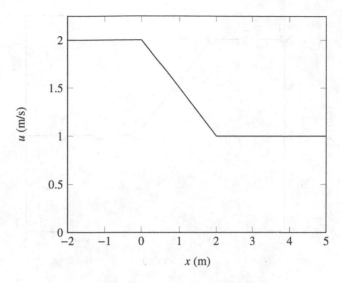

Figure 2.7 Initial Velocity distribution.

solution. This is not an ideal situation for our purposes because it adds an extra complication. However, instead of focusing on Equation 2.22, which uses a Eulerian description to obtain a solution for u for a given x and t, we will utilize characteristic curves to help decipher what the u distributions will look like at a certain time. To do so, we can will take advantage of the Lagrangian description of the Burgers' equation:

$$\frac{Du}{Dt} = 0.$$

To illustrate this approach to the Burgers' equation, consider the following example:

Example A system has an initial x-velocity distribution (u) given by:

$$u(x,0) = \begin{cases} 2 & x \leq 0 \\ -0.5x + 2 & 0 < x < 2 \\ 1 & x \geq 2 \end{cases}.$$

Figure 2.7 provides an illustration of the initial velocity distribution (in m/s) when x is between -2 m and 5 m. Plot the velocity distribution obtained by the Burgers' equation at $t = 0.5$ s, $t = 1$ s, $t = 1.5$ s, and $t = 2$ s. In addition, plot the characteristic curves for fluid elements whose initial positions are: $x = -2$ m, $x = -1$ m, $x = 0$ m, $x = 1$ m, $x = 2$ m, $x = 3$ m, $x = 4$ m.

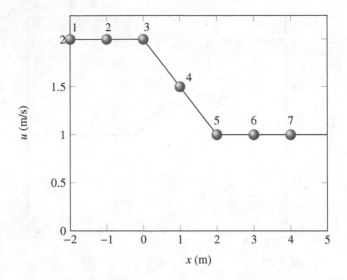

Figure 2.8 Marking fluid elements on the initial velocity distribution.

Solution:

As mentioned, we will use the Lagrangian description of the inviscid Burgers' equation to help us with this problem:

$$\frac{Du}{Dt} = 0$$

To help utilize this equation, we will need to mark fluid elements, as done in Figure 2.8, and track them as they travel. The velocity of these fluid elements will not change since the material derivative of velocity is zero. As in the advection equation example, the position of each of the fluid elements is given by the following expression:

$$x = x_0 + ut$$

where u is the velocity of each of the individual fluid elements and x_0 is the initial position. The velocity of each fluid element will remain constant, but the velocity may differ between different fluid elements. By tracking the fluid elements marked in Figure 2.8, we can map out the trajectory of each fluid element.

Table 2.2 provides values for the positions of each of the marked fluid elements at various times. This table will be used to determine the velocity distributions (i.e., the u vs x curve) at various times.

Using Table 2.2, we can plot the velocity distribution for the various time points, as given in Figure 2.9. By doing so, we notice that the high-velocity

Table 2.2 *Table of x and u values for the inviscid Burgers' equation example*

Fluid element	initial position (m)	u (m/s)	x (m) @t = 0.5s	x (m) @t = 1s	x (m) @t = 1.5s	x (m) @t = 2s
1	−2	2	−1	0	1	2
2	−1	2	0	1	2	3
3	0	2	1	2	3	4
4	1	1.5	1.75	2.5	3.25	4
5	2	1	2.5	3	3.5	4
6	3	1	3.5	4	4.5	5
7	4	1	4.5	5	5.5	6

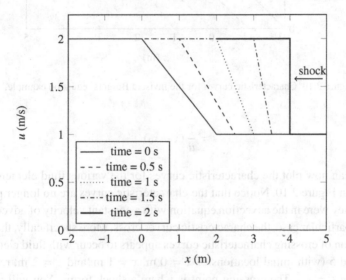

Figure 2.9 Plot of the velocity distribution.

fluid elements located on the left side of the initial velocity distribution begin to "catch up" to the slower moving fluid elements located on the right side of the initial distribution. As a result, the line in the middle becomes more vertical until eventually there is a sharp discontinuity (at $t = 2$ s) where the velocity will suddenly jump from a high value to a low value. This sudden jump is called a shock wave.

Drawing characteristic curves can be a useful tool in helping predict whether a shock forms. Recall that the characteristic curves are time versus position curves for each of the fluid elements. Thus, like before, we can rearrange $x = x_0 + ut$ in terms of t versus x to get:

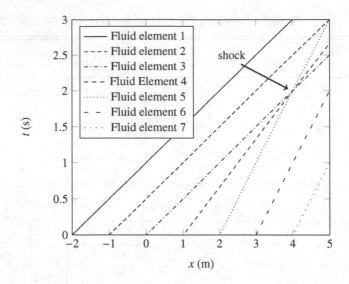

Figure 2.10 Characteristic curves for the inviscid Burgers' equation example.

$$t = \frac{1}{u}(x - x_0)$$

We can now plot the characteristic curves for the various fluid elements, as shown in Figure 2.10. Notice that the characteristic curves are no longer parallel, as they were in the advection equation with a constant velocity of advection. In this particular case, the characteristic curves cross. More specifically, the first indication of crossing characteristic curves appears to occur with fluid elements 3, 4, and 5 (with initial locations of $x = 0$ m, $x = 1$ m, and $x = 2$ m, respectively) at $t = 2$ s. The crossing point is where a shock forms. You will notice that the crossing point occurs at an x value of 4 m and a time of 2 s, which is the exact location and time of the shock wave given in Figure 2.9.

The formation of shock waves can be hindered somewhat by the inclusion of diffusion. Diffusion will be a topic we take up in Chapter 5. In short, diffusion acts somewhat as a "smoothing" agent.

It should be noted that our current approach to the Burgers' equation has limitations. For example, if we were to continue past $t = 2$ s using our current approach, the faster fluid elements would move even further past the slower moving fluid elements and a "Z" type of profile would begin to emerge. This is not physically possible since a "Z" profile would indicate multiple velocity values at a given x point. Thus, an alternative approach to capturing how the shock moves once it forms would be needed at that point. Nevertheless, the approach taken here provides the needed insight to understand the material derivative.

2.3 The Material Derivative and the Continuity Equation

We have already seen two versions of the continuity equation, namely an integral version (Equation 1.20):

$$\iiint_{\mathcal{V}} \frac{\partial \rho}{\partial t} d\mathcal{V} + \oiint_A \rho \vec{V} \cdot \vec{n} dA = 0$$

and a partial differential equation version known as the conservation form (Equation 1.23):

$$\frac{\partial \rho}{\partial t} + \vec{\nabla} \cdot \left(\rho \vec{V} \right) = 0.$$

We can rearrange the conservation form in terms of a material derivative. To do so, it might be easiest to start off by working in Cartesian coordinates, that is, Equation 1.24, rewritten here:

$$\frac{\partial \rho}{\partial t} + \frac{\partial (\rho u)}{\partial x} + \frac{\partial (\rho v)}{\partial y} + \frac{\partial (\rho w)}{\partial z} = 0.$$

Using the product rule of calculus, the spatial derivatives can be broken out via (note, del notation is underneath):

$$\frac{\partial \rho}{\partial t} + \underbrace{u \frac{\partial \rho}{\partial x} + v \frac{\partial \rho}{\partial y} + w \frac{\partial \rho}{\partial z}}_{\vec{V} \cdot \vec{\nabla} \rho} + \underbrace{\rho \frac{\partial u}{\partial x} + \rho \frac{\partial v}{\partial y} + \rho \frac{\partial w}{\partial z}}_{\rho \vec{\nabla} \cdot \vec{V}} = 0. \qquad (2.23)$$

The coordinate-free notation (i.e., using the del operator) of Equation 2.23 is written below:

$$\boxed{\frac{\partial \rho}{\partial t} + \vec{V} \cdot \vec{\nabla} \rho + \rho \vec{\nabla} \cdot \vec{V} = 0} \qquad (2.24)$$

Equation 2.24 is called the **non-conservation form of the continuity equation**. The first two terms of Equation 2.24 can be grouped into a material derivative of density. Thus, Equation 2.24 can be written as:

$$\boxed{\frac{D\rho}{Dt} + \rho \vec{\nabla} \cdot \vec{V} = 0} \qquad (2.25)$$

Equation 2.25 is called the **Lagrangian form of the continuity equation**. Equation 2.25 implies that the change of density of the fluid element as it travels

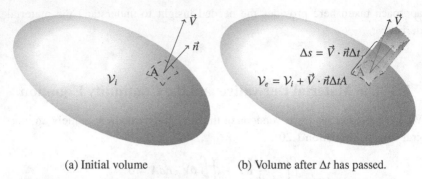

(a) Initial volume   (b) Volume after Δt has passed.

Figure 2.11 The volume of a fluid element (modeled as an ellipse) changes as it moves from one point to another in a given time, Δt. The velocity ($\vec{V}$) of a particular area segment with surface area, A, is shown. This particular area segment would "move away" from its original position, sweeping out a volume of $\vec{V} \cdot \vec{n}\Delta tA$. If only this area segment moved, the ending volume of the fluid element would be: $\mathcal{V}_e = \mathcal{V}_i + \vec{V} \cdot \vec{n}\Delta tA$. If all area segments moved, the ending volume would be: $\mathcal{V}_e = \mathcal{V}_i + \sum_{j=1}^{N} \vec{V}_j \cdot \vec{n}_j\Delta tA_j$, where N is the total number of area segments the volume is broken up into.

from one location to another is proportional (with ρ being the proportionality constant) to the negative of the divergence of velocity, i.e.:

$$\frac{D\rho}{Dt} = -\rho\vec{\nabla} \cdot \vec{V} \tag{2.26}$$

As you know, the density of an object is defined as its mass divided by its volume. Since the mass of the fluid element does not change, the only way for its density to change is if the volume changes. This fact would seem to indicate that the divergence of velocity is related to the volume of the fluid element.

To find out how the fluid element volume and the divergence of velocity are related, consider the following scenario depicted in Figure 2.11. This illustrates an arbitrary fluid element (which we will just take to be our ellipsoid "potato," except this time it is a fluid element and not a control volume) starting with an initial volume of $\mathcal{V}_i$ and changes to an ending volume of $\mathcal{V}_e$ in a time increment of Δt. In this situation, we are only changing the volume by considering a small area segment has moved out away from the original volume. To calculate the difference in the volume between the ending and initial states, we would need to calculate how much the area segment is displaced relative to its initial position. The displacement created by the area segment "sweeps" a volume. The volume is calculated by multiplying the normal distance the area segment has traveled away from the initial volume by its surface area, A. The displacement in the normal direction (Δs) can be written as the velocity normal to the surface multiplied by Δt:

$$\Delta s = \vec{V} \cdot \vec{n}\Delta t \tag{2.27}$$

where $\vec{n}$ is the outward unit normal vector pointing away from the fluid element and $\vec{V}$ is the velocity of the area segment. It is important to note in this situation that the velocity vector now describes the velocity of the area segment and not the velocity of the flow passing through the area segment (as was the case in the last chapter with the Eulerian approach). The velocity "dotted" with the unit normal vector gives the velocity component in the normal direction. With this definition for the displacement distance, the ending volume is just the initial volume plus the sum of the volume swept by the displacement of the area segment:

$$\mathcal{V}_e = \mathcal{V}_i + \underbrace{\vec{V} \cdot \vec{n}\Delta t}_{\Delta s} A \quad \text{(for one area segment)}.$$

If the volume is broken up into N area segments, then the ending volume is calculated by summing over all of the swept volumes of each area segment and adding it to the initial volume:

$$\mathcal{V}_e = \mathcal{V}_i + \sum_{j=1}^{N} \vec{V}_j \cdot \vec{n}_j \Delta t A_j \tag{2.28}$$

where the subscript j denotes the values for the jth area segment.

Rearranging Equation 2.28 by subtracting $\mathcal{V}_i$ and dividing by Δt leads to:

$$\frac{\mathcal{V}_e - \mathcal{V}_i}{\Delta t} = \sum_{j=1}^{N} \vec{V}_j \cdot \vec{n}_j A_j. \tag{2.29}$$

Taking the limits as $\Delta t \rightarrow 0$, $A_j \rightarrow dA$ (infinitesimal area), and $N \rightarrow \infty$ changes the left-hand side to the material derivative of volume and the right-hand side to a closed area integral:

$$\frac{D\mathcal{V}}{Dt} = \oiint_A \vec{V} \cdot \vec{n}dA. \tag{2.30}$$

Equation 2.30, which is an area integral, can be transformed into a volume integral using the divergence theorem. The divergence theorem changes the right-hand side of Equation 2.30 from an area integral to a volume integral via:

$$\oiint_A \vec{V} \cdot \vec{n}dA = \iiint_\mathcal{V} \vec{\nabla} \cdot \vec{V}d\mathcal{V}. \tag{2.31}$$

Plugging Equation 2.31 into Equation 2.30 leads to:

$$\frac{D\mathcal{V}}{Dt} = \iiint_\mathcal{V} \vec{\nabla} \cdot \vec{V}d\mathcal{V} \tag{2.32}$$

We now have a nice equation relating the material derivative of the volume of the fluid element to a volume integral of the divergence of velocity (i.e., $\vec{\nabla} \cdot \vec{V}$). We are going to cheat a little bit, mathematically speaking, and "shrink" down the volume in Equation 2.32 to an infinitesimal volume $(d\mathcal{V})$, to get:[5]

$$\frac{D(d\mathcal{V})}{Dt} = \vec{\nabla} \cdot \vec{V} d\mathcal{V}. \tag{2.33}$$

We can now relate the material derivative of density with the material derivative of volume by relating Equation 2.26 with Equation 2.33:

$$\frac{D\rho}{Dt} = -\frac{\rho}{d\mathcal{V}} \frac{D(d\mathcal{V})}{Dt}. \tag{2.34}$$

Some points should now be made here regarding the material derivative of volume. In particular note, from either Equation 2.32 or Equation 2.33, that if the material derivative of volume is zero, then the divergence of velocity $(\vec{\nabla} \cdot \vec{V})$ will be zero (which is the same as saying that the velocity field is divergence-free or solenoidal). Recall from the last chapter, when viewing from an Eulerian perspective, that a divergence-free velocity vector field implied "what goes in must come out" of the control volume. In a Lagrangian sense, a divergence-free velocity field implies that the volume of the fluid elements making up the flow is constant. We now come to a more general definition of an incompressible flow:

An incompressible flow is one where the volume of the moving fluid elements making up the flow do not change.

There are a couple of nice aspects of this definition of incompressible flow. One aspect is that it does not rely on any conservation principle (such as mass conversation). Instead, we obtained Equations 2.32 and 2.33 from a purely kinematic (motion only) argument. Another nice aspect of this definition is that it does not require the assumption that the density be the same value throughout the flow. This definition of incompressible flow only requires that the volume of each fluid element stay the same as it moves.

One other thing to note is that a consequence of an incompressible flow defined as the material derivative of volume is zero is that the mass conservation in a Langrangian description (Equation 2.25) simplifies to:

$$\frac{D\rho}{Dt} = 0.$$

You may also see this equation as the definition of an incompressible flow. That is, an incompressible flow is one where the density of the moving fluid elements making up the flow do not change. This definition is really just an

---

[5] In doing so, ridding ourselves of the volume integral.

extension of the definition of the volume of the fluid elements are constant since the mass of a fluid element stays fixed.

We will come across one more definition of an incompressible flow that deals with the speed of sound of a fluid in Section 6.6.

2.4 The Material Derivative in the Navier–Stokes Equations

One of the most common forms of the Navier–Stokes equations stems from applying Newton's laws of motion, in particular Newton's second law of motion, to a moving fluid element.

On many occasions Newton's second law is written as the sum (Σ) of forces equals mass times acceleration ($\Sigma \vec{F} = m\vec{a}$). However, we are going to utilize the more general way to write Newton's second law. We are going to write that force is equal to the time derivative of momentum (which is mass times velocity, $m\vec{V}$), i.e., $\Sigma \vec{F} = \frac{d(m\vec{V})}{dt}$. The mass is the density multiplied by the volume of the fluid element. We made an assumption earlier when we developed the material derivative that a fluid element would have well-defined property values (i.e., a well-defined density, temperature, etc.). We are going to extend our definition of a fluid element slightly to now allow for varying properties, such as density, within each fluid element. This will allow us to do what we did with mass conservation in Chapter 1 and write the momentum as a volume integral. Thus, our force balance (Newton's second Law) for a fluid element will look like this:

$$\frac{d}{dt}\underbrace{\left(\overbrace{\iiint_{\mathcal{V}(t)} \rho \vec{V} d\mathcal{V}}^{m\vec{V}} \right)}_{\frac{d(m\vec{V})}{dt}} = \Sigma \vec{F}. \qquad (2.35)$$

Note that we are writing the summation of forces on the right and the time derivative of momentum on the left (as opposed to the other way around), this order is the typical convention in fluid mechanics. Notice that the term on the left-hand side is essentially a mass term (i.e., density, ρ, times volume, $d\mathcal{V}$) multiplied by a velocity term, $\vec{V}$. Thus, the integral indicates the momentum of a fluid element whose density (and velocity) may vary within the volume of the fluid element. In addition, since we are using the Lagrangian description, the volume integral is integrated over a volume whose value is dependent on time since the volume of a fluid element can change over time. We have now encountered a problem. The issue here is that we want to take the time derivative of an integral whose limits are a function of time (i.e., integrating over $\mathcal{V}(t)$).

There is a formal procedure for taking the derivative of an integral with varying limits,[6] For our purposes we are going to come up with a "cheat" to take the derivative of this integral. Suppose you wanted to sneak the derivative inside the integral. By doing so, the ordinary derivative needs to change because ρ and the $\vec{V}$ vary not only with time, but also with space (x, y, z). Since this is a Lagrangrian description, we should change our ordinary derivative to a material derivative, i.e.:

$$\underbrace{\iiint_{\mathcal{V}(t)} \frac{D\left(\rho\vec{V}d\mathcal{V}\right)}{Dt}}_{\frac{d}{dt}\left(\iiint_{\mathcal{V}(t)} \rho\vec{V}d\mathcal{V}\right)} = \Sigma\vec{F}. \qquad (2.36)$$

Using the product rule for the material derivative the same way you would when taking an ordinary derivative would yield:

$$\iiint_{\mathcal{V}(t)} \left(\frac{D\rho}{Dt}\vec{V}d\mathcal{V} + \rho\frac{D\vec{V}}{Dt}d\mathcal{V} + \rho\vec{V}\frac{D(d\mathcal{V})}{Dt}\right) = \Sigma\vec{F}.$$

We can flip the second and third terms on the left-hand side to get:

$$\iiint_{\mathcal{V}(t)} \left(\frac{D\rho}{Dt}\vec{V}d\mathcal{V} + \rho\vec{V}\frac{D(d\mathcal{V})}{Dt} + \rho\frac{D\vec{V}}{Dt}d\mathcal{V}\right) = \Sigma\vec{F}.$$

We can now use Equation 2.33 $\left(\frac{D(d\mathcal{V})}{Dt} = \vec{\nabla} \cdot \vec{V}d\mathcal{V}\right)$ to relate the material derivative of the infinitesimal volume (i.e., the second term on the left-hand side of the equal sign) to the divergence of velocity to get:

$$\iiint_{\mathcal{V}(t)} \left(\frac{D\rho}{Dt}\vec{V}d\mathcal{V} + \rho\vec{V}\vec{\nabla} \cdot \vec{V}d\mathcal{V} + \rho\frac{D\vec{V}}{Dt}d\mathcal{V}\right) = \Sigma\vec{F}.$$

Factoring out $\vec{V}d\mathcal{V}$ from the first and second term and distributing the integral through leads to:

$$\iiint_{\mathcal{V}(t)} \underbrace{\left(\frac{D\rho}{Dt} + \rho\vec{\nabla} \cdot \vec{V}\right)}_{\text{look familiar?}}\vec{V}d\mathcal{V} + \iiint_{\mathcal{V}(t)} \rho\frac{D\vec{V}}{Dt}d\mathcal{V} = \Sigma\vec{F}.$$

The integrand in the first integral in the equation above is zero due to the mass conservation equation in Lagrangian form, Equation 2.25 $\left(\frac{D\rho}{Dt} + \rho\vec{\nabla} \cdot \vec{V} = 0\right)$. Thus, we now have:

---

[6] The mathematical procedure for taking a derivative of an integral with varying limits is called differentiation under the integral sign, or the Leibniz rule. In fluid mechanics, the Leibniz rule is generalized in a theorem called the Reynolds transport theorem.

$$\underbrace{\iiint_{\mathcal{V}(t)} \rho \frac{D\vec{V}}{Dt} d\mathcal{V} = \Sigma \vec{F}}_{\substack{\textbf{will be a useful starting point} \\ \textbf{for the Navier–Stokes equations}}} . \qquad (2.37)$$

It seems as though a lot of work was done to get a result we may have expected already. That is, the left-hand side is nothing but a mass term (i.e., density times volume, $\rho d\mathcal{V}$) multiplied by the acceleration (i.e., the derivative of velocity with respect to time, $\vec{a} = \frac{D\vec{V}}{Dt}$). Nevertheless, going through this whole process is a worthwhile exercise. Equation 2.37 will be a useful starting point when we develop the Navier–Stokes equations in the next chapter.

2.5 Take Home Points

In this chapter, we discussed the two approaches to fluid flow: the Lagrangian and Eulerian descriptions. We studied the advection equation and the inviscid Burgers' equation. We considered mass conservation from a Lagrangian perspective (unlike Chapter 1 where we only considered mass from an Eulerian perspective). In addition, we considered a force balance on a moving fluid, giving us a starting point for developing the Navier–Stokes equations in the next chapter.

Along the way, you may have noticed that there are multiple forms the conservation principles can take. In particular, these forms are:

- **Integral form**: Stems directly from applying a conservation principle in an Eulerian description to a finite-sized control volume.
- **Conservation form**: Can be obtained by "shrinking" down the finite-sized control volume in the integral form to an infinitesimally sized control volume.
- **Lagrangian form**: Stems directly from applying a conservation principle to a moving fluid element in a Lagrangian description.
- **Non-conservation form**: Can be obtained by expanding out the material derivative in the Lagrangian form to an Eulerian description.

Chapter 1 considered mass conservation using an Eulerian description and developed equations for the integral form and conservation form. Section 2.3 discussed the Lagrangian form and non-conservation form of mass conservation, i.e., Equations 2.25 and 2.24, respectively. Although we did not do this, we could have derived Equations 2.25 and 2.24 by applying mass conservation to a moving fluid.[7]

In Section 2.4, the time rate of change of momentum in Newton's second law was taken on a moving fluid element. This yielded Equation 2.37, which will be a starting point for the development of the Navier–Stokes equations.

In the next chapter, we will discuss the various forces that act on a fluid element, eventually leading to the development of the Lagrangian and non-conservation form of the Navier–Stokes equations.

Problems

2.1   Given an initial temperature distribution of $T(x) = \sin(x)$, create a table similar to Table 2.1 for times of 0 s, 0.5 s, 1 s, and 1.5 s when $u = 1$ m/s. Plot the final temperature distribution at $t = 1.5$ s.

2.2   Explain why the characteristic curves for the advection equation are parallel when the velocity is a constant value.

2.3   Given an initial x-velocity distribution of:

$$u(x,0) = \begin{cases} 1 & x \le 0.5 \\ 2x & 0.5 < x < 1. \\ 2 & x \ge 1 \end{cases}$$

Does a shock form when the inviscid Burgers' equation is applied to this initial velocity field? Why or why not? Plot the characteristic curves.

2.4   Find an expression for the material derivative of temperature if the temperature is given by the following equation $T = e^{-t}(\sin(2x) + \cos(y))$ and if the velocity vector is: $\vec{V} = \hat{i} + 2\hat{j}$.

2.5   For the temperature equation used in Problem 2.4, what is the advective transport term of temperature if the velocity field is given by $\vec{V} = 2xy\hat{i} - y^2\hat{j}$ when $x = 2$, $y = 1$, and $t = 1$?

2.6   Given the velocity field vector: $\vec{V} = y\,(A\cos(2t) + B\sin(3t))\,\hat{i} + 6xyt\hat{j}$ m/s, what is the acceleration of the fluid at $x = 0.5$ m and $y = 0.5$ m at time, $t = 1$ s?

---

[7] Since the mass of a fluid element does not change, the mass conservation in a Lagrangian description would simply be the time derivative of the mass of a moving fluid element is equal to zero, that is:

$$\frac{d}{dt} \underbrace{\iiint_{\mathcal{V}(t)} \rho d\mathcal{V}}_{\text{mass of fluid element}} = 0$$

2.7 The Lagrangian and non-conservation form of the continuity equation (i.e., Equations 2.25 and 2.24) can be obtained by applying mass conservation to a moving fluid element, which states that the mass of a moving fluid element does not change in time. Mathematically, this can be written as:

$$\frac{d}{dt} \iiint_{V(t)} \rho d\mathcal{V} = 0$$

Using the ideas from Section 2.4, where we applied a time derivative to the momentum of a moving fluid element, derive the Lagrangian form of the mass continuity equation as well as the non-conservation form of the continuity equation.

2.8 Suppose the density of a fluid element is given by the expression: $\rho = e^{-0.005t} + 1$. It travels with a velocity of 2 m/s in the x-direction. What is the value of the material derivative after it has gone 60 meters?

2.9 What is the summation force vector on a cube fluid element that is 1 meter by 1 meter by 1 meter in size if the density is 1 kg/m^3 and $\vec{V} = 2x\hat{i} + 4y\hat{j}$?

2.10 The gradient of a scalar function, f, is defined as:

$$\vec{\nabla}f \equiv \hat{i}\frac{\partial f}{\partial x} + \hat{j}\frac{\partial f}{\partial y} + \hat{k}\frac{\partial f}{\partial z}.$$

Show that $\left(\vec{V} \cdot \vec{\nabla}\right)f = \vec{V} \cdot \left(\vec{\nabla}f\right)$

3

Force Balance, the Stress Tensor, and the Navier–Stokes Equations

In the last two chapters, we have used both a Lagrangian point of view and an Eulerian point of view to develop different, but equivalent, versions of the continuity equation. The Eulerian point of view dominated our discussion in Chapter 1 and the Lagrangian point of view was the focus of Chapter 2.

Like the continuity equation, the Navier–Stokes equations can be obtained either through a Lagrangian description or an Eulerian description. At the end of Chapter 2, we developed a starting point for developing the Navier–Stokes equations by considering Newton's second law on a moving fluid element. We developed the equation:

$$\iiint_{\mathcal{V}(t)} \rho \frac{D\vec{V}}{Dt} d\mathcal{V} = \Sigma \vec{F} \tag{2.37}$$

This equation is using a Lagrangian description, which is the most natural and most common description when discussing Newton's laws of motion. This starting point will naturally lead to the Lagrangian form of the Navier–Stokes equations.

3.1 Forces on a Fluid and the Stress Tensor

In general, forces acting on a fluid element can be categorized into two groups: body forces ($\vec{F}_b$) and surface forces ($\vec{F}_{surf}$). Thus, the summation of forces ($\Sigma \vec{F}$) can be written as:

$$\Sigma \vec{F} = \vec{F}_{surf} + \vec{F}_b \tag{3.1}$$

A **body force** is a force that acts on all parts of the fluid element, including the internal parts of the fluid element. The easiest body force to visualize and the one we will use in this book is a gravitational force. A gravitational force is a body force because all internal parts of the fluid element (such as the atoms that make up the fluid element) experience gravity.

A simple expression of the gravitational body force acting on a mass, m, is the mass times the gravitational acceleration ($\vec{g}$), leading to $\vec{F}_b = m\vec{g}$. If the density of the fluid element (ρ) and the gravitational acceleration ($\vec{g}$) are both constant, then we can write the body force as $\vec{F}_b = \rho \mathcal{V} \vec{g}$, where $\mathcal{V}$ is the volume of the fluid element and ρ is the density (and the mass is $\rho \mathcal{V}$). However, this idea can be generalized to deal with a non-uniform density and gravitational acceleration within the fluid element. This leads to a more general form of the gravitational body force using a volume integral:

$$\vec{F}_b = \iiint_{\mathcal{V}(t)} \rho \vec{g} d\mathcal{V} \tag{3.2}$$

Notice that we are integrating over a volume that is a function of time since we are assuming that we are following a moving fluid element whose volume could change.

The other type of force is the surface force. They are called **surface forces** because they stem from surface to surface contact with another body (such as other fluid elements) and are not a force that is applied to the internal guts of the fluid element (such as a body force). A surface force is usually written in terms of a stress term multiplied by an area. The stress term we will utilize is called the stress vector (or traction vector) and is given by the symbol $\vec{\tau}$.[1] Figure 3.1 illustrates the stress vector acting on a small surface segment of the fluid element. The surface segment shown has a differential area of dA. The **stress vector (traction vector)** illustrated in Figure 3.1 is just the force per unit area on the surface segment shown, with units of newton per meter squared (or pascal, Pa). Thus, a differential surface force, $d\vec{F}_{surf}$, acting on the surface segment of the fluid element is just the stress vector multiplied by the area, i.e., $d\vec{F}_{surf} = \vec{\tau} dA$. Integrating over the entire surface (i.e., summing up all of the little differential force calculations over the entire surface of the fluid element) leads to the following expression for the surface force:

$$\vec{F}_{surf} = \oiint_{A(t)} \vec{\tau} dA \tag{3.3}$$

Note that there is no explicit reference to the orientation of the differential area element that a stress vector might be acting on in Equation 3.3. In other words, there is no outward unit normal used in Equation 3.3. This leads to the question of how do we account for the different orientations of each differential area. After all, stress acting in the x-direction on a differential surface oriented

---

[1] The traction vector is also sometimes denoted using $\vec{\sigma}$ instead of $\vec{\tau}$.

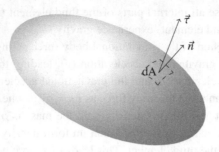

Figure 3.1 Illustration of stress vector.

in one direction will have a different effect on the fluid element than the same stress acting on a differential surface oriented in a different direction. In order to take into account both of these directions, that is, the stress vector direction and the orientation of the differential surface element, we will need to introduce a new concept. That concept is the stress tensor, which we will denote with the symbol, $\overset{\leftrightarrow}{T}$.

The **stress tensor** is the link between two vectors: the stress vector and the vector associated with the orientation of the area segment (i.e., the outward unit normal). Considering that the stress vector and the outward unit normal do not necessarily have to align themselves, the stress tensor must encode information about the direction of each of them. The fact that the stress tensor contains information about two different directions (one for the stress vector and one for the outward normal vector) provides a clue as to what is called the order of the tensor. The stress tensor, since it encodes information from two different vectors, is considered a second-order tensor. Note that vectors such as the velocity vector are considered first-order tensors and scalars, which have no directional dependence such as temperature, are called zero-order tensors. The double arrow above the $\overset{\leftrightarrow}{T}$ used for the stress tensor indicates a second-order tensor. A common property of a tensor is that they are invariant under a coordinate transformation. In other words, a tensor encodes the same information whether you are in Cartesian or cylindrical coordinates. As an example, the velocity vector (a first-order tensor) "points" in the same direction and has the same magnitude no matter if it is represented in Cartesian coordinates, cylindrical coordinates, or some other coordinate system. The components of the velocity vector in Cartesian coordinates or cylindrical coordinates will be different, but the vector itself is left unchanged in different coordinate representations.

The stress vector is related to the stress tensor via a dot product of the stress tensor with the outward unit normal, $\vec{n}$:[2]

$$\vec{\tau} = \vec{\vec{T}} \cdot \vec{n} \tag{3.4}$$

The best way to get a handle on Equation 3.4 and the stress tensor in general is to see what the stress tensor looks like in a particular coordinate system. For simplicity, we will look at the stress tensor in Cartesian coordinates. The stress tensor in Cartesian coordinates can be represented as a matrix as follows:

$$\vec{\vec{T}} = \begin{pmatrix} \tau_{xx} & \tau_{xy} & \tau_{xz} \\ \tau_{yx} & \tau_{yy} & \tau_{yz} \\ \tau_{zx} & \tau_{zy} & \tau_{zz} \end{pmatrix} \tag{3.5}$$

The way to think about the components of the stress tensor is the following: each term has two subscripts associated with a direction. The first subscript denotes the direction of a particular "face" on a Cartesian element and the second subscript denotes the direction of a stress acting on that face. In other words:

The faces (denoted by the first subscript in the stress tensor components) are determined by the direction of the outward unit normal. For example, τ_{xy} is a stress in the y-direction acting on a face whose normal is in the x-direction (either the positive or negative x-direction). Thus, we say that τ_{xy} is a stress on the x-face in the y-direction. Similarly, we say that τ_{zx} is the stress on the z-face in the x-direction, τ_{yz} is the stress on the y-face in the z-direction, and so on. Upon inspection of the stress tensor in Equation 3.5, you might recognize that the columns are the forces (i.e., stresses) in a particular direction and the rows are the forces (stresses) acting on a particular face.

Figure 3.2 provides the stresses acting on all faces, with the shaded faces being the x-faces. Figure 3.2a shows the stresses acting on the positive x-, y-, and z-faces and Figure 3.2b shows the stresses acting on the negative x-, y-, and z-faces. If we were to "shrink" the Cartesian element down to an infinitesimal volume, the stresses on the negative faces should be equal and opposite the stresses on the positive faces due to Newton's third law of motion. To illustrate this point, consider a simpler scenario such as that given in Figure 3.3. For this

---

[2] In some texts, Equation 3.4 is written as:

$$\vec{\tau} = \vec{n} \cdot \vec{\vec{T}}$$

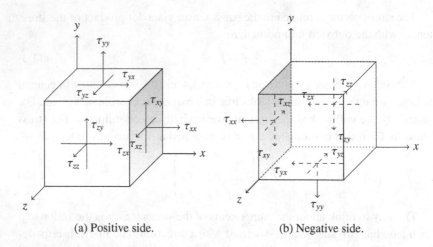

(a) Positive side.   (b) Negative side.

Figure 3.2 Illustration of the stress tensor in Cartesian coordinates. The shaded faces are the x-faces since the normal of those faces is in the x-direction. The faces with positive normals are shown in (a) and the faces with negative normals are shown in (b). Incidentally, the faces whose normals are in the negative direction also have forces acting in the negative directions.

fluid element, the only stress present is the stress acting on the x-face in the x-direction, i.e., the τ_{xx} component of the stress tensor. Notice in Figure 3.3 that we are distinguishing the stress on the left side (i.e., the negative x-face) as $\tau_{xx,l}$ with the stress on the right side (i.e., the positive x-face) as $\tau_{xx,r}$. If the fluid element is infinitesimal in size in the x-dimension, then the "positive" x-face and the "negative" x-face come together as one surface. If the stress on the "positive" side goes to the right (as it is doing in Figure 3.3), then a stress that is equal in magnitude but opposite in direction needs to be applied to the other side (i.e., the "negative" side) of the infinitesimally thin surface. In other words, $\tau_{xx,r} = -\tau_{xx,l}$. Thus, for an infinitesimally small element, the stresses acting on a positive face need to have partners that are equal and opposite on the corresponding negative-direction faces. Note that this does not mean that the stresses acting on the positive faces will always be positive and that the stresses acting on the negative faces will always be negative (as appears to be the case in Figure 3.2). Instead, it just means that, as the Cartesian element shrinks to an infinitesimal size, the stresses on the positive and negative faces need to be in the opposite direction. The way in which the stress tensor encapsulates this

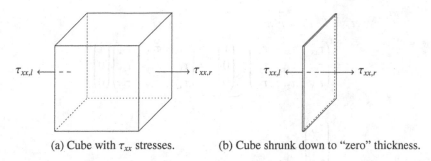

(a) Cube with τ_{xx} stresses.   (b) Cube shrunk down to "zero" thickness.

Figure 3.3 Demonstration of Newton's Third Law.

information becomes more apparent when we take the dot product of the stress tensor with the unit normal.

Recall from Chapter 1 that the dot product of two vectors performs a projection of one vector onto another. The dot product of the second-order stress tensor with the unit normal vector is also doing a form of projection. The projection this dot product is performing is a projection of the "stress state" of a fluid element onto a particular surface of that element defined by its outward unit normal vector. The stress tensor is essentially encoding all of the stress information of the fluid element in one mathematical object. You might be wondering how we "dot" a second-order tensor ($\vec{\vec{T}}$) with a vector ($\vec{n}$) considering the standard definition of a dot product involves two vectors, resulting in a scalar. As discussed in Chapter 1, the easiest way to take the dot product between two vectors, in Cartesian coordinates, is to matrix multiply the transpose of one vector with the other vector. Similarly, taking the dot product of our second-order stress tensor (i.e., $\vec{\vec{T}}$) with a vector (i.e., $\vec{n}$) amounts to nothing but simple matrix multiplication of the transpose of the tensor with the vector,[3] leading to another vector (e.g., $\vec{\tau}$). As an example, suppose we take the unit normal to be in the positive x-direction, that is $\vec{n} = \hat{\imath}$ (or the "right" face), then we get the following result for $\vec{\vec{T}} \cdot \vec{n}$ (with $\vec{n}$ written as a column vector):

---

[3] Just to emphasize, this approach is particularly useful for Cartesian coordinates. Care must be taken when dealing with non-orthogonal coordinates. Such situations do arise in fluid mechanics, however, this would require diving into the weeds of tensor analysis and would divert attention away from the current analysis.

$$\vec{\tau}_r = \vec{\vec{T}} \cdot \vec{n}_r = \vec{\vec{T}}^\dagger \vec{n}_r$$

$$= \begin{pmatrix} \tau_{xx} & \tau_{xy} & \tau_{xz} \\ \tau_{yx} & \tau_{yy} & \tau_{yz} \\ \tau_{zx} & \tau_{zy} & \tau_{zz} \end{pmatrix}^\dagger \underbrace{\begin{pmatrix} 1 \\ 0 \\ 0 \end{pmatrix}}_{\vec{n}=\hat{\imath}} = \begin{pmatrix} \tau_{xx} & \tau_{yx} & \tau_{zx} \\ \tau_{xy} & \tau_{yy} & \tau_{zy} \\ \tau_{xz} & \tau_{yz} & \tau_{zz} \end{pmatrix} \begin{pmatrix} 1 \\ 0 \\ 0 \end{pmatrix}$$

$$= \begin{pmatrix} \tau_{xx} \\ \tau_{xy} \\ \tau_{xz} \end{pmatrix}_r,$$

where we are using the subscript 'r' in $\vec{\tau}_r$ to just denote the stresses on the right-hand face of a Cartesian element. Recall that the transpose flips the rows and columns and is denoted with a $\dagger$ superscript. Notice that when we take the stress tensor and we "dot" it with a unit normal vector, we get the stress vector acting on the surface associated with the unit normal. In this case, we get the stress vector acting on the positive x-face, as shown in Figure 3.2a.

Similarly, we can take the dot product with the negative x-face, i.e., with $\vec{n} = -\hat{\imath}$ (the "left" face):

$$\vec{\tau}_l = \vec{\vec{T}} \cdot \vec{n}_l = \vec{\vec{T}}^\dagger \vec{n}_l$$

$$= \begin{pmatrix} \tau_{xx} & \tau_{xy} & \tau_{xz} \\ \tau_{yx} & \tau_{yy} & \tau_{yz} \\ \tau_{zx} & \tau_{zy} & \tau_{zz} \end{pmatrix}^\dagger \underbrace{\begin{pmatrix} -1 \\ 0 \\ 0 \end{pmatrix}}_{\vec{n}=-\hat{\imath}} = \begin{pmatrix} \tau_{xx} & \tau_{yx} & \tau_{zx} \\ \tau_{xy} & \tau_{yy} & \tau_{zy} \\ \tau_{xz} & \tau_{yz} & \tau_{zz} \end{pmatrix} \begin{pmatrix} -1 \\ 0 \\ 0 \end{pmatrix}$$

$$= - \begin{pmatrix} \tau_{xx} \\ \tau_{xy} \\ \tau_{xz} \end{pmatrix}_l$$

As you can see, the stress on the negative x-face is the negative of the stress on the positive x-face. Thus, the direction of the stress vector on the negative x-face is the opposite of the direction of the stress vector acting on the positive x-face. The stress acting on the negative x-face is given in Figure 3.2b. There are two points to recognize here:

- The stresses on the negative and positive face have the same magnitude (but different direction) when the Cartesian element is of infinitesimal size (i.e., when the stresses are evaluated at effectively the same point in space). However, it very well could be that the stress on a finite-sized Cartesian element will be different on the negative faces than on the positive faces if the element is big enough.

- The stresses on the positive face will not always be in the positive direction and the stresses on the negative face will not always be in the negative direction. For example, we could have drawn the arrows to be flipped in Figure 3.3, making $\tau_{xx,r}$ in the negative direction on the positive face and $\tau_{xx,l}$ positive on the negative face. In that scenario, the two stresses would be pointing inward towards each other, "squishing" the element. This would be considered a compressive stress and the value of τ_{xx} in the stress tensor matrix would be negative.

Speaking of compression stresses, some additional definitions might be in order here: The terms along the diagonal (e.g., τ_{xx}, τ_{yy}, τ_{zz}) of the stress tensor are called **normal stresses** and the terms on the off diagonal (e.g., τ_{xy}, τ_{zy}, τ_{yx}, etc.) are sometimes called **shear stresses**. Normal stress components that are positive are considered **tensile stresses**, whereas normal stress components that are negative are **compressive stresses**.

3.1.1 Angular Momentum and Stress Tensor Symmetry

We have seen how the stresses on the negative faces are equal and opposite (in direction) to the stresses on the positive faces when the element is shrunk down to an infinitesimal size. This is essentially satisfying Newton's third law of motion. At this point, however, we have not discussed any potential rotation of the element. To consider rotation, we are going to look at a simple example, as illustrated in Figure 3.4. In Figure 3.4, we are only concerned with shear stresses on the x- and y-faces. In addition, we are going to ignore any z-direction stress. Notice we are distinguishing the stresses on the faces on the left, right, top, and bottom using subscripts l, r, t, and b, respectively. As the Cartesian element shrinks to an infinitesimal size, we know that $\tau_{xy,r} = -\tau_{xy,l}$ and $\tau_{yx,t} = -\tau_{yx,b}$. In addition, the resistance to any rotation goes to zero (i.e., the moment of inertia goes to zero). Thus any net torque that would cause a rotation on an infinitesimally sized element would cause the angular acceleration of the element to approach infinity (since there would be no moment of inertia to resist such rotation). Therefore, as the element shrinks down, the net torque acting on the element needs to go to zero. Since the torque is caused by the shear stresses acting on the fluid element, there needs to be a balance between these stresses in order for there to be a net zero torque. Examining Figure 3.4, we note that the stresses $\tau_{xy,r}$ and $\tau_{xy,l}$ appears to rotate the element counter-clockwise and the stresses $\tau_{yx,t}$ and $\tau_{yx,b}$ appear to rotate the element in a clockwise fashion. Given this observation and the fact that we cannot have a net torque on the

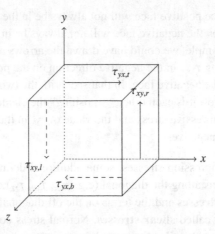

Figure 3.4 Shearing stresses that can cause a rotation.

Cartesian element if it is of an infinitesimal size, we can make the following deduction:

$$\tau_{xy,r} = \tau_{yx,t} \quad \text{(positive faces)}$$

and

$$\tau_{xy,l} = \tau_{yx,b} \quad \text{(negative faces)}$$

Since at an infinitesimal size the magnitudes of $\tau_{xy,r}$ and $\tau_{xy,l}$ are the same, and the magnitudes of $\tau_{yx,t}$ and $\tau_{yz,b}$ are the same, we can reach the following conclusion:

$$\tau_{xy} = \tau_{yx}$$

A similar thought process would show that:

$$\tau_{xz} = \tau_{zx}$$

and

$$\tau_{yz} = \tau_{zy}$$

Thus, it turns out, that the stress tensor is symmetric.[4] A more formal procedure illustrating the symmetry of the stress tensor can be done using conservation of angular momentum.

---

[4] This assumes that the fluid element has no internal source of torque, which can happen (such as when there are electrostatic body forces).

3.1.2 Total Force Expression

Now that we have discussed some of the basic ideas of the stress tensor, we can plug the relationship between the stress vector and stress tensor, i.e., Equation 3.4 or $\vec{\tau} = \vec{\vec{T}} \cdot \vec{n}$, into our surface force equation, i.e., Equation 3.3 or $\vec{F}_{surf} = \oiint_{A(t)} \vec{\tau} dA$, to get:

$$\vec{F}_{surf} = \underbrace{\oiint_{A(t)} \vec{\vec{T}} \cdot \vec{n} dA}_{\text{can use divergence theorem}} \tag{3.6}$$

Notice that Equation 3.6 is an area integral that now has a dot product with the unit normal vector, i.e., there is a $\cdot \vec{n} dA$ term. This is an indication that we can utilize the divergence theorem again. Using the divergence theorem on Equation 3.6 yields a volume integral with the integrand being the divergence of the stress tensor:

$$\vec{F}_{surf} = \iiint_{V(t)} \vec{\nabla} \cdot \vec{\vec{T}} dV \tag{3.7}$$

Thus, our summation of forces, i.e., Equation 3.1 or $\Sigma \vec{F} = \vec{F}_{surf} + \vec{F}_b$, can be written in one of two ways:

$$\Sigma \vec{F} = \oiint_{A(t)} \vec{\vec{T}} \cdot \vec{n} dA + \iiint_{V(t)} \rho \vec{g} dV \tag{3.8}$$

or

$$\Sigma \vec{F} = \iiint_{V(t)} \vec{\nabla} \cdot \vec{\vec{T}} dV + \iiint_{V(t)} \rho \vec{g} dV \tag{3.9}$$

Equation 3.8 is the result of summing the surface force via Equation 3.6 with the gravitational body force, i.e., Equation 3.2 or $\vec{F}_b = \iiint_{V(t)} \rho \vec{g} dV$, and Equation 3.9 is the result of summing the surface force via Equation 3.7 with the body force. We will use both Equation 3.8 and Equation 3.9 in developing the various forms of the Navier–Stokes equations.

3.2 General Force Balance: Cauchy's First Law of Motion

We have discussed the various forces acting on a general element. We broke the forces into surface and body forces as well as discussed the idea of a stress vector (traction vector) and stress tensor. We can introduce the force expressions we obtained in the previous section as the next big step toward the Navier–Stokes equations.

We begin by inserting Equation 3.9 into the general force balance, i.e., Equation 2.37 $\left(\iiint_{\mathcal{V}(t)} \rho \frac{D\vec{V}}{Dt} d\mathcal{V} = \Sigma \vec{F} \right)$, to get:

$$\iiint_{\mathcal{V}(t)} \rho \frac{D\vec{V}}{Dt} d\mathcal{V} = \iiint_{\mathcal{V}(t)} \vec{\nabla} \cdot \vec{\vec{T}} d\mathcal{V} + \iiint_{\mathcal{V}(t)} \rho \vec{g} d\mathcal{V} \qquad (3.10)$$

Shrinking down to an infinitesimal volume would yield:

$$\rho \frac{D\vec{V}}{Dt} d\mathcal{V} = \vec{\nabla} \cdot \vec{\vec{T}} d\mathcal{V} + \rho \vec{g} d\mathcal{V}$$

which would give us a final result of a general force balance on a fluid element (after dividing out the $d\mathcal{V}$):

$$\boxed{\rho \frac{D\vec{V}}{Dt} = \vec{\nabla} \cdot \vec{\vec{T}} + \rho \vec{g}} \qquad (3.11)$$

Equation 3.11 is a very important result. It is sometimes called Cauchy's equation of motion or Cauchy's first law of motion. In particular, Equation 3.11 is the **Lagrangian form of Cauchy's first law of motion.**[5]

Cauchy's equation of motion is a general equation and applies to all materials (whether they be solids or fluids), provided that the material can still be modeled as a continuum. What distinguishes the material used in the calculation is the form that the stress tensor takes. Determining an expression for the stress tensor for given materials is a big part of continuum mechanics and materials science research. Such stress tensor studies are often referred to as developing a constitutive model for a given material.

3.2.1 Divergence of the Stress Tensor

You may have noticed that Equation 3.11 contains a divergence of the stress tensor ($\vec{\nabla} \cdot \vec{\vec{T}}$). You might also be wondering how to take the divergence of a second-order tensor.

Ultimately, the divergence of a second-order tensor is really nothing but taking the divergence of each column (vector) in the stress tensor. To perform such an operation in Cartesian coordinates, we can somewhat think of it like matrix multiplication except we are now taking derivatives instead of multiplying (much like we did when we evaluated $\vec{\nabla} \cdot \vec{V}$ in Cartesian coordinates in Chapter 1). Thus, the divergence of the stress tensor term, in Cartesian coordinates, becomes:

---

[5] Equation 3.11 is also sometimes called Cauchy's momentum equation. Note, in some texts, you may see Cauchy's first law written in terms of a static situation, i.e., with the left-hand side equaling zero.

$$\vec{\nabla} \cdot \vec{\vec{T}} = \begin{pmatrix} \frac{\partial}{\partial x} & \frac{\partial}{\partial y} & \frac{\partial}{\partial z} \end{pmatrix} \cdot \begin{pmatrix} \tau_{xx} & \tau_{xy} & \tau_{xy} \\ \tau_{yx} & \tau_{yy} & \tau_{yz} \\ \tau_{zx} & \tau_{zx} & \tau_{zz} \end{pmatrix} \tag{3.12}$$

$$= \begin{pmatrix} \frac{\partial \tau_{xx}}{\partial x} + \frac{\partial \tau_{yx}}{\partial y} + \frac{\partial \tau_{zx}}{\partial z}, & \frac{\partial \tau_{xy}}{\partial x} + \frac{\partial \tau_{yy}}{\partial y} + \frac{\partial \tau_{zy}}{\partial z}, & \frac{\partial \tau_{xz}}{\partial x} + \frac{\partial \tau_{yz}}{\partial y} + \frac{\partial \tau_{zz}}{\partial z} \end{pmatrix}$$

Notice that the divergence of the stress tensor leads to a row vector. However, as we shall see later, it will be more convenient to write the result of this divergence as a column vector. Thus, taking the transpose of Equation 3.12 becomes:

$$\vec{\nabla} \cdot \vec{\vec{T}} = \begin{pmatrix} \frac{\partial \tau_{xx}}{\partial x} + \frac{\partial \tau_{yx}}{\partial y} + \frac{\partial \tau_{zx}}{\partial z} \\ \frac{\partial \tau_{xy}}{\partial x} + \frac{\partial \tau_{yy}}{\partial y} + \frac{\partial \tau_{zy}}{\partial z} \\ \frac{\partial \tau_{xz}}{\partial x} + \frac{\partial \tau_{yz}}{\partial y} + \frac{\partial \tau_{zz}}{\partial z} \end{pmatrix} \tag{3.13}$$

The fact that we are somewhat arbitrarily switching from a row vector to a column vector may alarm some, but it is easier to work with column vectors. In general, a notation called tensor notation (as opposed to the del notation we are using) is often employed, which avoids such arbitrary manipulation of the vectors from rows to columns. We are not going to discuss tensor notation but as you progress in your studies on fluid mechanics you may want to become familiar with the notation.

3.3 The Form of the Stress Tensor

Many fluids can be modeled by what is known as a Newtonian fluid. The Navier–Stokes equations typically assume a Newtonian fluid, so getting a handle on the Newtonian fluid stress tensor is important in understanding the Navier–Stokes equations. Instead of covering the formal procedure for obtaining the Newtonian fluid stress tensor, let's instead attempt to develop it through more informal means. In general, we can think of the Newtonian fluid stress tensor as having three terms, with each term corresponding to a force associated with the fluid. In words, the stress tensor can be broken up as:

$$\vec{\vec{T}} = \begin{matrix} \text{force (per area)} \\ \text{on a fluid} \\ \text{element when} \\ \text{there is no} \\ \text{motion} \\ \text{(first term)} \end{matrix} \cdot + \begin{matrix} \text{force (per area)} \\ \text{on a fluid} \\ \text{element due to} \\ \text{expansion or} \\ \text{contraction} \\ \text{(second term)} \end{matrix} + \begin{matrix} \text{force (per area)} \\ \text{on a fluid} \\ \text{element due to} \\ \text{shearing} \\ \text{(third term)} \end{matrix}$$

The terms of the stress tensor can be mapped to various fluid mechanics concepts as follows:

- the first term → associated with pressure (p)
- the second term → associated with the divergence of velocity ($\vec{\nabla} \cdot \vec{V}$)
- the third term → associated with the gradient of velocity ($\vec{\nabla}\vec{V}$)

Each term has units of pascal (Pa), which is a newton per meter squared ($\frac{N}{m^2}$). Let's examine each term.

First term in the stress tensor – pressure

The first term in the stress tensor consists of a force (per area) that would exist on our fluid element even if there is no fluid motion. This results in a pressure (p) term, as in the thermodynamic pressure acting on the walls of a container filled with air. The origin of this pressure is the atoms bouncing into each other as well as into the walls of the container. Since the stress tensor is written as a matrix, but the pressure is only a scalar quantity, the pressure quantity is multiplied by an identity matrix ($\vec{I}$) which would put the pressure term along the diagonal of a matrix. Thus, the first term can be written as:

$$firstTerm = -p\vec{I}$$

In matrix form, this would look like:

$$-p\vec{I} = -p \begin{pmatrix} 1 & 0 & 0 \\ 0 & 1 & 0 \\ 0 & 0 & 1 \end{pmatrix} = \begin{pmatrix} -p & 0 & 0 \\ 0 & -p & 0 \\ 0 & 0 & -p \end{pmatrix} \tag{3.14}$$

The first thing to note about our pressure term is that we have included a minus sign. We will find that the negative sign on the pressure term is needed in order for our fluid velocity calculations to make physical sense. The negative sign means that a positive pressure value ($p > 0$) indicates a compressive force. Thus, when drawing pressure on a Cartesian fluid element, as done in Figure 3.5, the direction of the arrows for pressure is always drawn inward. In addition, note that pressure has the same value at a given point no matter the direction a "measurement" is made (i.e., it is the same whether you look at an x-face, y-face, or a z-face). This independence of direction is called isotropic.

One more comment, at the moment we will consider pressure to be a thermodynamic variable but, as we shall see, in certain situations (such as in incompressible flows) the pressure will not be thermodynamic in nature but instead will have more of a mechanical description.

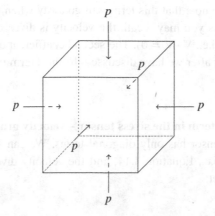

Figure 3.5 Illustrates the pressure acting on a Cartesian element. Notice how the pressure is conventionally drawn inward since it is considered positive when compressing a fluid element.

Second term in the stress tensor – divergence of velocity

The next term in our stress tensor deals with the force on a fluid element which would cause either a contraction or expansion in size. Thus, we can relate this force to the volume change of the fluid element, or more specifically, the material derivative of volume. You may recall that the material derivative of volume is related to the divergence of velocity (i.e., Equation 2.33 or $\frac{D(dV)}{Dt} = \vec{\nabla} \cdot \vec{V} dV$). Therefore, we can postulate that the second term will be proportional to the divergence of velocity (i.e., $\lambda \vec{\nabla} \cdot \vec{V}$, where λ is a proportionality constant). The name of the proportionality constant is often called the second coefficient of viscosity and has units of pascal-seconds, or Pa · s. Like pressure (p), the divergence of velocity is a scalar. Thus, the second term will need to be multiplied by the identity matrix in order to appropriately incorporate it in the stress tensor, i.e.:

$$secondTerm = \lambda\left(\vec{\nabla} \cdot \vec{V}\right)\vec{I}$$

We can write this term out in matrix form as:

$$\lambda\left(\vec{\nabla} \cdot \vec{V}\right)\vec{I} = \lambda\left(\vec{\nabla} \cdot \vec{V}\right)\begin{pmatrix} 1 & 0 & 0 \\ 0 & 1 & 0 \\ 0 & 0 & 1 \end{pmatrix}$$

$$= \begin{pmatrix} \lambda\left(\frac{\partial u}{\partial x} + \frac{\partial v}{\partial y} + \frac{\partial w}{\partial z}\right) & 0 & 0 \\ 0 & \lambda\left(\frac{\partial u}{\partial x} + \frac{\partial v}{\partial y} + \frac{\partial w}{\partial z}\right) & 0 \\ 0 & 0 & \lambda\left(\frac{\partial u}{\partial x} + \frac{\partial v}{\partial y} + \frac{\partial w}{\partial z}\right) \end{pmatrix} \quad (3.15)$$

You may want to note that this term will go away when the flow is incompressible because, as you may recall, the velocity is divergence-free in an incompressible flow (i.e., $\vec{\nabla} \cdot \vec{V} = 0$). The second coefficient of viscosity, λ, will be discussed further after we have discussed the third term in our stress tensor.

Third term in the stress tensor – velocity gradient

So far, our stress tensor has only diagonal terms. We can see this by adding the pressure term, i.e., Equation 3.14, and the velocity divergence term, i.e., Equation 3.15, to get:

$$\vec{\vec{T}} = firstTerm + secondTerm + thirdTerm$$

$$= \begin{pmatrix} -p + \lambda\vec{\nabla} \cdot \vec{V} & 0 & 0 \\ 0 & -p + \lambda\vec{\nabla} \cdot \vec{V} & 0 \\ 0 & 0 & -p + \lambda\vec{\nabla} \cdot \vec{V} \end{pmatrix} + thirdTerm$$

The last term (i.e., the *thirdTerm*) will "fill" in the off diagonal locations of the stress tensor, in addition to adding some contribution to the diagonal. Ultimately, the last term in the stress tensor deals with a force due to shearing the fluid element. This shearing will cause the velocity of various parts of the fluid element to spatially vary. For instance, in a simple scenario, consider the situation in Figure 3.6. This figure illustrates a Cartesian fluid element (just shown in two dimensions for clarity). The top of the fluid element (i.e., the positive y-face) is given a force in the x-direction. This force (or stress) is a shear stress, denoted by τ_{yx}. With this applied stress, the top begins to move at a velocity of u. In this example, the bottom of the fluid element is considered to be held fixed with zero velocity. A stress is considered a **shear stress** when the direction of the stress is parallel to the surface face, or in other words, is perpendicular to the normal of the face. You can think of this stress much like you would a friction force.

This stress can be related to the change of velocity with respect to space, i.e., x-, y-, and z-. So, for example, in the scenario given in Figure 3.6, the u velocity (i.e., the x-velocity) changes with respect to the up and down direction (i.e., the y-direction). The velocity difference between the top and bottom surfaces will cause a deformation of the fluid element, as shown in Figure 3.6. It turns out, for a Newtonian fluid, that the stress applied to the top surface of the fluid element (i.e., the τ_{yx} stress) can be modeled as having a linear relationship with

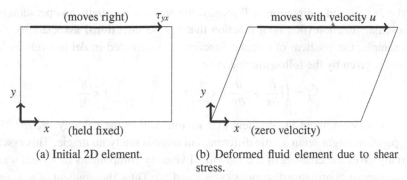

(a) Initial 2D element.   (b) Deformed fluid element due to shear stress.

Figure 3.6 Two-dimensional fluid element with an applied shear stress at the top face, i.e., $\vec{\tau} = \tau_{yx}\hat{i}$ applied to the surface defined by $\vec{n} = \hat{j}$.

the velocity change (i.e., derivative) in the y-direction. In mathematical terms, the expression for τ_{yx} for a Newtonian fluid can be given by:

$$\tau_{yx} = \mu\frac{\partial u}{\partial y} \qquad (3.16)$$

where μ is a material property known as the dynamic viscosity and has units of pascal-second (Pa·s). The higher the dynamic viscosity, the more resistance to motion the fluid will be, i.e., the more "viscous." To give you some numbers, at 20°C, water has a dynamic viscosity of about 1×10^{-3} Pa·s, air has a dynamic viscosity of 1.8×10^{-5} Pa·s, and honey has a dynamic viscosity of 8 Pa·s.

Not all materials or fluids will have such a nice relationship between the shear stress and the velocity derivative. However, many fluids (such as air and water) can be modeled using this simple model. As mentioned, this model is called a Newtonian fluid. A **Newtonian fluid** is one where the dynamic viscosity is independent of how much shear stress is applied to the fluid element. Non-Newtonian fluids, such as ketchup, toothpaste, and corn starch in water, may have varying dynamic viscosities depending on the stress level. Thus, for some non-Newtonian fluids, the more you stress it, the even greater the resistance will be, and visa-versa.

Equation 3.16 is just one stress direction (i.e., the x-direction) acting on one face (i.e., the y-face). There are also x- and z-faces as well as y- and z-stress directions. Thus, in general, we need some way to take into account multiple directions of stress being applied to multiple faces. We can encapsulate these various directions and faces into one entity known as the velocity gradient ($\vec{\nabla}\vec{V}$). You might be wondering how to take a gradient of the velocity vector. As a reminder, the gradient is a three-dimensional analog of a derivative of a function. It provides the change of some function, f, in the various directions (i.e., x-, y-, and z-directions). Typically, the gradient is usually an operation for a scalar function (that is, a function that has no directional association). For example, the gradient of a scalar function, f, is denoted in del notation as $\vec{\nabla}f$ and is given by the following operation:

$$\vec{\nabla}f = \left(\hat{i}\frac{\partial}{\partial x} + \hat{j}\frac{\partial}{\partial y} + \hat{k}\frac{\partial}{\partial z}\right)f = \hat{i}\frac{\partial f}{\partial x} + \hat{j}\frac{\partial f}{\partial y} + \hat{k}\frac{\partial f}{\partial z}$$

Now, what about the gradient of a vector, such as the velocity vector? The operation might seem a little different, but there is really no magic. This essentially gives the changes of the individual velocity components (u, v, and w) in the various coordinate directions (x-, y-, and z-). Thus, the gradient of u, v, and w is respectively:

$$\vec{\nabla}u = \left(\hat{i}\frac{\partial}{\partial x} + \hat{j}\frac{\partial}{\partial y} + \hat{k}\frac{\partial}{\partial z}\right)u = \hat{i}\frac{\partial u}{\partial x} + \hat{j}\frac{\partial u}{\partial y} + \hat{k}\frac{\partial u}{\partial z} \tag{3.17}$$

$$\vec{\nabla}v = \left(\hat{i}\frac{\partial}{\partial x} + \hat{j}\frac{\partial}{\partial y} + \hat{k}\frac{\partial}{\partial z}\right)v = \hat{i}\frac{\partial v}{\partial x} + \hat{j}\frac{\partial v}{\partial y} + \hat{k}\frac{\partial v}{\partial z} \tag{3.18}$$

$$\vec{\nabla}w = \left(\hat{i}\frac{\partial}{\partial x} + \hat{j}\frac{\partial}{\partial y} + \hat{k}\frac{\partial}{\partial z}\right)w = \hat{i}\frac{\partial w}{\partial x} + \hat{j}\frac{\partial w}{\partial y} + \hat{k}\frac{\partial w}{\partial z} \tag{3.19}$$

We can "group" the gradients of the velocity components into one matrix by writing each gradient as a column vector, thus the velocity gradient ($\vec{\nabla}\vec{V}$) becomes:

$$\vec{\nabla}\vec{V} = \begin{pmatrix} \frac{\partial u}{\partial x} & \frac{\partial v}{\partial x} & \frac{\partial w}{\partial x} \\ \frac{\partial u}{\partial y} & \frac{\partial v}{\partial y} & \frac{\partial w}{\partial y} \\ \frac{\partial u}{\partial z} & \frac{\partial v}{\partial z} & \frac{\partial w}{\partial z} \end{pmatrix} \tag{3.20}$$

The result given in 3.20 is the velocity gradient in Cartesian coordinates. This is just one of a number of ways to obtain the result given in 3.20.[6]

The velocity gradient is considered a 2nd order tensor. Let's take stock of what we are doing here. We are trying to come up with a relationship for the third term in the stress tensor, which we presumed would deal with the shearing of a fluid element. We made the claim that, for a simple system given in Figure 3.6, the shear stress on the y-face in the x-direction can be given by: $\tau_{yx} = \mu\frac{\partial u}{\partial y}$. You may now be tempted to think that we can state a general form for the

---

[6] One way, which we won't cover in this text, to compute the velocity gradient in Cartesian coordinates is by writing the del operator, $\vec{\nabla}$, in column form and the velocity vector as a row vector (as opposed to column vector like usual) and perform what amounts to an "outer" product type operation.

third term to just be: *thirdTerm* $= \mu \vec{\nabla}\vec{V}$. This would put the stress tensor as (by adding Equations 3.14, 3.15, and 3.20):

$$\vec{\vec{T}} = firstTerm + secondTerm + thirdTerm$$

$$= \begin{pmatrix} -p + \lambda\vec{\nabla}\cdot\vec{V} + \mu\frac{\partial u}{\partial x} & \mu\frac{\partial v}{\partial x} & \mu\frac{\partial w}{\partial x} \\ \mu\frac{\partial u}{\partial y} & -p + \lambda\vec{\nabla}\cdot\vec{V} + \mu\frac{\partial v}{\partial y} & \mu\frac{\partial w}{\partial y} \\ \mu\frac{\partial u}{\partial z} & \mu\frac{\partial v}{\partial z} & -p + \lambda\vec{\nabla}\cdot\vec{V} + \mu\frac{\partial w}{\partial z} \end{pmatrix}$$

$$\underbrace{}$$

NOTE: This is NOT correct

However, this is not quite right. The reason is that the form for the stress tensor given above is not symmetric and recall from Section 3.1.1 that the stress tensor needs to be symmetric. As an example, τ_{yx} should equal τ_{xy}. However, in the proposed stress tensor above, they do not equal each other as $\tau_{yx} = \mu\frac{\partial u}{\partial y}$ and $\tau_{xy} = \mu\frac{\partial v}{\partial x}$. A simple proposal to "work around" the symmetry problem is to write the third term as a sum of the velocity gradient and the transpose of the velocity gradient. Thus, a new proposal for our third term will be:

$$thirdTerm = \mu\left(\vec{\nabla}\vec{V} + \left(\vec{\nabla}\vec{V}\right)^{\dagger}\right)$$

Writing out our third term in Cartesian form gives us:

$$\mu\left(\vec{\nabla}\vec{V} + \left(\vec{\nabla}\vec{V}\right)^{\dagger}\right) = \mu\left(\begin{pmatrix} \frac{\partial u}{\partial x} & \frac{\partial v}{\partial x} & \frac{\partial w}{\partial x} \\ \frac{\partial u}{\partial y} & \frac{\partial v}{\partial y} & \frac{\partial w}{\partial y} \\ \frac{\partial u}{\partial z} & \frac{\partial v}{\partial z} & \frac{\partial w}{\partial z} \end{pmatrix} + \begin{pmatrix} \frac{\partial u}{\partial x} & \frac{\partial u}{\partial y} & \frac{\partial u}{\partial z} \\ \frac{\partial v}{\partial x} & \frac{\partial v}{\partial y} & \frac{\partial v}{\partial z} \\ \frac{\partial w}{\partial x} & \frac{\partial w}{\partial y} & \frac{\partial w}{\partial z} \end{pmatrix}\right)$$

$$(3.21)$$

$$= \begin{pmatrix} 2\mu\frac{\partial u}{\partial x} & \mu\left(\frac{\partial v}{\partial x} + \frac{\partial u}{\partial y}\right) & \mu\left(\frac{\partial w}{\partial x} + \frac{\partial u}{\partial z}\right) \\ \mu\left(\frac{\partial u}{\partial y} + \frac{\partial v}{\partial x}\right) & 2\mu\frac{\partial v}{\partial y} & \mu\left(\frac{\partial w}{\partial y} + \frac{\partial v}{\partial z}\right) \\ \mu\left(\frac{\partial u}{\partial z} + \frac{\partial w}{\partial x}\right) & \mu\left(\frac{\partial v}{\partial z} + \frac{\partial w}{\partial y}\right) & 2\mu\frac{\partial w}{\partial z} \end{pmatrix}$$

Thus, we have a symmetric tensor we can use. Notice that the third term has diagonal terms as well as off-diagonal terms. Thus, this term is technically not just shear stresses, even though we are implying that it is only a shear stress term.

Final form of Newtonian Fluid Stress Tensor

We can add Equations 3.14, 3.15, and 3.21 together to get:

$$\vec{\vec{T}} = -p\vec{\vec{I}} + \lambda\left(\vec{\nabla}\cdot\vec{V}\right)\vec{\vec{I}} + \mu\left(\vec{\nabla}\vec{V} + \left(\vec{\nabla}\vec{V}\right)^{\dagger}\right)$$

$$= \begin{pmatrix} -p+\lambda\left(\vec{\nabla}\cdot\vec{V}\right)+2\mu\frac{\partial u}{\partial x} & \mu\left(\frac{\partial v}{\partial x}+\frac{\partial u}{\partial y}\right) & \mu\left(\frac{\partial w}{\partial x}+\frac{\partial u}{\partial z}\right) \\ \\ \mu\left(\frac{\partial u}{\partial y}+\frac{\partial v}{\partial x}\right) & -p+\lambda\left(\vec{\nabla}\cdot\vec{V}\right)+2\mu\frac{\partial v}{\partial y} & \mu\left(\frac{\partial w}{\partial y}+\frac{\partial v}{\partial z}\right) \\ \\ \mu\left(\frac{\partial u}{\partial z}+\frac{\partial w}{\partial x}\right) & \mu\left(\frac{\partial v}{\partial z}+\frac{\partial w}{\partial y}\right) & -p+\lambda\left(\vec{\nabla}\cdot\vec{V}\right)+2\mu\frac{\partial w}{\partial z} \end{pmatrix}$$

$$(3.22)$$

We currently have two material parameters in Equation 3.22, the dynamic viscosity (μ) and the second coefficient of viscosity (λ). The dynamic viscosity can be obtained from experiments but the second coefficient of viscosity is much more elusive to determine empirically. However, George Gabriel Stokes in the mid-1800s proposed a relationship between the dynamic viscosity and the second coefficient of viscosity. The relationship he proposed was the following:

$$\lambda = -\frac{2}{3}\mu \qquad (3.23)$$

Stokes obtained this relationship by first defining a new kind of pressure, one that includes both the pressure we defined earlier (p) as well as the additional terms along the diagonals of the stress tensor. The new type of pressure is called mechanical pressure, or mean pressure. The mechanical pressure is essentially obtained by taking the average of the diagonal terms of the stress tensor. Thus we need to divide the trace[7] of the stress tensor by 3 (actually, we will divide by negative three, you will see why shortly). Thus, from Equation 3.22, the mechanical pressure (p_m) will be:

$$p_m = -\frac{1}{3}trace\left(\vec{\vec{T}}\right)$$

$$= -\frac{1}{3}\left(-p+\lambda\left(\vec{\nabla}\cdot\vec{V}\right)+2\mu\frac{\partial u}{\partial x}-p+\lambda\left(\vec{\nabla}\cdot\vec{V}\right)+2\mu\frac{\partial v}{\partial y}-p+\lambda\left(\vec{\nabla}\cdot\vec{V}\right)+2\mu\frac{\partial w}{\partial z}\right)$$

$$= -\frac{1}{3}\left(-3p+3\lambda\vec{\nabla}\cdot\vec{V}+2\mu\frac{\partial u}{\partial x}+2\mu\frac{\partial v}{\partial y}+2\mu\frac{\partial w}{\partial z}\right)$$

$$= p - \lambda\vec{\nabla}\cdot\vec{V} - \frac{2}{3}\mu\underbrace{\left(\frac{\partial u}{\partial x}+\frac{\partial v}{\partial y}+\frac{\partial w}{\partial z}\right)}_{\vec{\nabla}\cdot\vec{V}} \qquad (3.24)$$

---

[7] Recall that the trace of a matrix is just the summation of the diagonal elements.

So the mechanical (or mean) pressure is:

$$p_m = p - \lambda \vec{\nabla} \cdot \vec{V} - \frac{2}{3}\mu \vec{\nabla} \cdot \vec{V} \qquad (3.25)$$

Stokes proposed the expression for the second coefficient of viscosity such that the mechanical pressure would resort back to the pressure we defined earlier (p). Note as well, if we did not divide by negative three but instead positive three, the mechanical pressure would be defined as the negative of p.

Plugging Equation 3.23 for the second coefficient of viscosity into Equation 3.22 results in the well-known **Newtonian fluid stress tensor**:

$$\boxed{\vec{\vec{T}} = -p\vec{\vec{I}} - \frac{2}{3}\mu\left(\vec{\nabla} \cdot \vec{V}\right)\vec{\vec{I}} + \mu\left(\vec{\nabla}\vec{V} + \left(\vec{\nabla}\vec{V}\right)^{\dagger}\right)} \qquad (3.26)$$

We can now write the stress tensor in Cartesian coordinates using Equations 3.14, 3.15, and 3.21, via (with $\lambda = -\frac{2}{3}\mu$):

$$\vec{\vec{T}} = -p\vec{\vec{I}} - \frac{2}{3}\mu\left(\vec{\nabla} \cdot \vec{V}\right)\vec{\vec{I}} + \mu\left(\vec{\nabla}\vec{V} + \left(\vec{\nabla}\vec{V}\right)^{\dagger}\right) =$$

$$\begin{pmatrix} -p - \frac{2}{3}\mu\left(\vec{\nabla} \cdot \vec{V}\right) + 2\mu\frac{\partial u}{\partial x} & \mu\left(\frac{\partial v}{\partial x} + \frac{\partial u}{\partial y}\right) & \mu\left(\frac{\partial w}{\partial x} + \frac{\partial u}{\partial z}\right) \\ \mu\left(\frac{\partial u}{\partial y} + \frac{\partial v}{\partial x}\right) & -p - \frac{2}{3}\mu\left(\vec{\nabla} \cdot \vec{V}\right) + 2\mu\frac{\partial v}{\partial y} & \mu\left(\frac{\partial w}{\partial y} + \frac{\partial v}{\partial z}\right) \\ \mu\left(\frac{\partial u}{\partial z} + \frac{\partial w}{\partial x}\right) & \mu\left(\frac{\partial v}{\partial z} + \frac{\partial w}{\partial y}\right) & -p - \frac{2}{3}\mu\left(\vec{\nabla} \cdot \vec{V}\right) + 2\mu\frac{\partial w}{\partial z} \end{pmatrix}$$

$$\qquad (3.27)$$

Therefore, each of the individual stress tensor components can be written in Cartesian coordinates as:

$$\tau_{xx} = -p - \frac{2}{3}\mu\left(\vec{\nabla} \cdot \vec{V}\right) + 2\mu\frac{\partial u}{\partial x}$$

$$\tau_{yy} = -p - \frac{2}{3}\mu\left(\vec{\nabla} \cdot \vec{V}\right) + 2\mu\frac{\partial v}{\partial y}$$

$$\tau_{zz} = -p - \frac{2}{3}\mu\left(\vec{\nabla} \cdot \vec{V}\right) + 2\mu\frac{\partial w}{\partial z}$$

$$\tau_{xy} = \tau_{yx} = \mu\left(\frac{\partial v}{\partial x} + \frac{\partial u}{\partial y}\right)$$

$$\tau_{xz} = \tau_{zx} = \mu\left(\frac{\partial w}{\partial x} + \frac{\partial u}{\partial z}\right)$$

$$\tau_{yz} = \tau_{zy} = \mu\left(\frac{\partial v}{\partial z} + \frac{\partial w}{\partial y}\right)$$

Again, notice the symmetry of the stress tensor. That is, we could transpose (or flip the rows and columns) and the result will look the same.

3.4 The Navier–Stokes Equations ... Finally

Equation 3.26 is the stress tensor we are going to use for developing the compressible Navier–Stokes equations. Plugging Equation 3.26 into Cauchy's first law (Equation 3.11, $\rho \frac{D\vec{V}}{Dt} = \vec{\nabla} \cdot \vec{\vec{T}} + \rho \vec{g}$) leads to:

$$\rho \frac{D\vec{V}}{Dt} = \vec{\nabla} \cdot \left(-p\vec{\vec{I}} - \frac{2}{3}\mu \left(\vec{\nabla} \cdot \vec{V} \right) \vec{\vec{I}} + \mu \left(\vec{\nabla}\vec{V} + \left(\vec{\nabla}\vec{V} \right)^{\dagger} \right) \right) + \rho \vec{g}$$

Distributing the divergence leads to:

$$\rho \frac{D\vec{V}}{Dt} = \vec{\nabla} \cdot \left(-p\vec{\vec{I}} \right) - \vec{\nabla} \cdot \left(\frac{2}{3}\mu \left(\vec{\nabla} \cdot \vec{V} \right) \vec{\vec{I}} \right) + \vec{\nabla} \cdot \left(\mu \left(\vec{\nabla}\vec{V} + \left(\vec{\nabla}\vec{V} \right)^{\dagger} \right) \right) + \rho \vec{g}$$

The divergence of the first two terms on the right side of the equal sign can be written as gradients (we will show this later) to yield the **compressible Navier–Stokes equations in Lagrangian form**:

$$\boxed{\rho \frac{D\vec{V}}{Dt} = -\vec{\nabla}p - \vec{\nabla} \left(\frac{2}{3}\mu \left(\vec{\nabla} \cdot \vec{V} \right) \right) + \vec{\nabla} \cdot \left(\mu \left(\vec{\nabla}\vec{V} + \left(\vec{\nabla}\vec{V} \right)^{\dagger} \right) \right) + \rho \vec{g}} \qquad (3.28)$$

We can expand out the material derivative of velocity in Equation 3.28 using Equation 2.12 developed in Chapter 2 to write the **non-conservation form of the compressible Navier–Stokes equations**.

$$\boxed{\rho \left(\frac{\partial \vec{V}}{\partial t} + \vec{V} \cdot \vec{\nabla}\vec{V} \right) = -\vec{\nabla}p - \vec{\nabla} \left(\frac{2}{3}\mu \left(\vec{\nabla} \cdot \vec{V} \right) \right) + \vec{\nabla} \cdot \left(\mu \left(\vec{\nabla}\vec{V} + \left(\vec{\nabla}\vec{V} \right)^{\dagger} \right) \right) + \rho \vec{g}}$$

$$(3.29)$$

Below are the compressible Navier–Stokes equations with common terms labeled:

Compressible Navier–Stokes equations (non-conservation form)

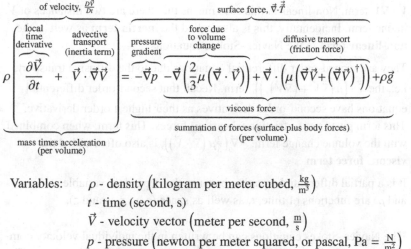

Variables: ρ - density $\left(\text{kilogram per meter cubed, } \frac{\text{kg}}{\text{m}^3}\right)$

t - time (second, s)

$\vec{V}$ - velocity vector $\left(\text{meter per second, } \frac{\text{m}}{\text{s}}\right)$

p - pressure $\left(\text{newton per meter squared, or pascal, Pa} = \frac{\text{N}}{\text{m}^2}\right)$

μ - dynamic viscosity (pascal-second, Pa·s)

$\vec{g}$ - gravitational acceleration $\left(\text{meters per second squared, } \frac{\text{m}}{\text{s}^2}\right)$

$\vec{\nabla}$ - del (gradient) operator $\left(\text{inverse meter, } \frac{1}{\text{m}}\right)$

$\vec{\nabla}\cdot$ - divergence operator $\left(\text{inverse meter, } \frac{1}{\text{m}}\right)$

$\vec{V}\cdot\nabla$ - advective term operator $\left(\text{inverse second, } \frac{1}{\text{s}}\right)$

The units of this equation[8] are newtons per volume, or $\frac{\text{N}}{\text{m}^3}$. The left-hand side of this version of the Navier–Stokes equations is just the mass (written in terms of density, or mass per volume) multiplied by the acceleration. The acceleration is the material derivative of velocity written out in an Eulerian description. The right-hand side is the summation of forces acting on the fluid, broken up into surface forces and a gravitational body force.

The various terms of the surface force will be discussed shortly. For now, we can make some general comments about this equation:

- The dominant variable in these equations is the velocity vector, $\vec{V}$. However, the pressure (p) is also considered an unknown, as is density (ρ). Note, for

---

[8] Note that the Navier–Stokes equations have been stated in the plural sense, that is, with the "s" at the end of the equations, even though it looks like it is just one equation. The reason for this is because the Navier–Stokes equations consist of three separate equations once broken out into components for a particular coordinate system.

an incompressible flow, the density is typically considered to be a known input value obtained from a lookup table.

- The equations are non-linear due to the advective transport term, i.e., the $\vec{V} \cdot \vec{\nabla} \vec{V}$ term. Non-linear in this case means that there are two instances of $\vec{V}$ in one term. Incidentally, this is also called the **inertia term** as well as the **non-linear term** in the Navier–Stokes equations.

- They are second-order differential equations due to the diffusive transport, i.e., the $\vec{\nabla} \cdot \left(\mu \left(\vec{\nabla} \vec{V} + \left(\vec{\nabla} \vec{V} \right)^{\dagger} \right) \right)$, term. Recall that second-order differential equations have second-order derivatives as their highest order derivative. This term gives rise to friction (shearing) forces. This term, when combined with the volume change term, $-\vec{\nabla} \left(\frac{2}{3} \mu \left(\vec{\nabla} \cdot \vec{V} \right) \right)$, is also often called the **viscous force term**.

- It is a partial differential equation because the dependent variables ($\vec{V}$, p, and ρ) are functions of time, t, as well as space (e.g., x, y, and z).

The Navier–Stokes equations can be written in the individual velocity components in the three spatial directions (e.g., in the x-, y-, and z-directions). The unknowns are the individual velocity components in each of the three spatial directions, as well as pressure, and density. This gives us five unknowns but only three equations. The continuity equation will provide an equation for density, rewritten here in conservation form:

$$\frac{\partial \rho}{\partial t} + \vec{\nabla} \cdot \left(\rho \vec{V} \right) = 0$$

We still need another equation. We can add an equation of state to the mix. As was discussed in Chapter 1, the equation of state relates pressure, temperature, and density. An example of an equation of state is the ideal gas equation of state:

$$p = \rho R T$$

where R is a gas constant specific to the gas used in the flow.

The equation of state provides an additional equation, giving us five equations: continuity, Navier–Stokes (three components), and equation of state. Unfortunately, we now have six unknowns: density, three components of velocity, pressure, and temperature. So, we need one last equation for temperature. We get this last equation for temperature through the conservation of energy equation, which we will discuss in Chapter 5. The conservation of energy equation will provide our sixth and last equation to close the set.

As you can probably tell by now, there is quite a bit going on. We are now going to discuss the individual terms that make up the surface force. Namely, the pressure gradient term, the force due to volume change term, and the diffu-sive transport term.

3.4.1 Pressure Gradient Term

The Navier–Stokes equations has a term coined the pressure gradient term. It is the boxed term below:

$$\rho\left(\frac{\partial \vec{V}}{\partial t} + \vec{V} \cdot \vec{\nabla}\vec{V}\right) = \boxed{-\vec{\nabla}p} - \vec{\nabla}\left(\frac{2}{3}\mu\left(\vec{\nabla}\cdot\vec{V}\right)\right) + \vec{\nabla}\cdot\left(\mu\left(\vec{\nabla}\vec{V} + \left(\vec{\nabla}\vec{V}\right)^{\dagger}\right)\right) + \rho\vec{g}$$

The pressure gradient term comes directly from applying the divergence of the Newtonian stress tensor, written here in full:

$$\vec{\nabla} \cdot \vec{T} = \vec{\nabla} \cdot \left(-p\vec{I} - \frac{2}{3}\mu\left(\vec{\nabla}\cdot\vec{V}\right)\vec{I} + \mu\left(\vec{\nabla}\vec{V} + \left(\vec{\nabla}\vec{V}\right)^{\dagger}\right)\right)$$

$$= \boxed{\vec{\nabla} \cdot \left(-p\vec{I}\right)} + \vec{\nabla}\cdot\left(-\frac{2}{3}\mu\left(\vec{\nabla}\cdot\vec{V}\right)\vec{I}\right) + \vec{\nabla}\cdot\left(\mu\left(\vec{\nabla}\vec{V} + \left(\vec{\nabla}\vec{V}\right)^{\dagger}\right)\right)$$

Thus, the divergence of the pressure term in our stress tensor $(-p\vec{I})$ leads to the pressure gradient term. We can take the divergence of this pressure term in Cartesian coordinates as follows:

$$\underbrace{\begin{pmatrix} \frac{\partial}{\partial x} & \frac{\partial}{\partial y} & \frac{\partial}{\partial z} \end{pmatrix} \begin{pmatrix} -p & 0 & 0 \\ 0 & -p & 0 \\ 0 & 0 & -p \end{pmatrix}}_{\vec{\nabla}\cdot\left(-p\vec{I}\right)} = \underbrace{-\begin{pmatrix} \frac{\partial p}{\partial x} & \frac{\partial p}{\partial y} & \frac{\partial p}{\partial z} \end{pmatrix}}_{\substack{-\vec{\nabla}p \\ \text{-(gradient of p)}}} \qquad (3.30)$$

As you can see, the divergence of $-p\vec{I}$ leads to the gradient of the scalar pressure function, p. For future convenience, we will write the gradient of pressure as a column vector as opposed to a row vector, therefore:

$$\vec{\nabla} \cdot \left(-p\vec{I}\right) = -\vec{\nabla}p = -\begin{pmatrix} \frac{\partial p}{\partial x} \\ \frac{\partial p}{\partial y} \\ \frac{\partial p}{\partial z} \end{pmatrix} \qquad (3.31)$$

What does the gradient of pressure mean, physically? The easiest way to understand the meaning behind the pressure gradient is to consider a simple fluid element, such as the square fluid element given in Figure 3.7. We will only consider pressure on the "x-"faces. We can write the pressure on the left edge

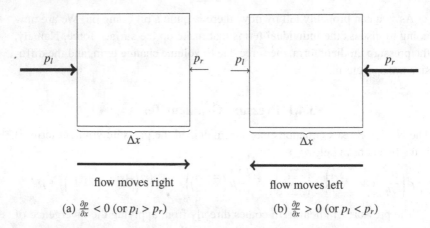

(a) $\frac{\partial p}{\partial x} < 0$ (or $p_l > p_r$)   (b) $\frac{\partial p}{\partial x} > 0$ (or $p_l < p_r$)

Figure 3.7 Pressure gradient in the x-direction on a 2D square fluid element. If the pressure on the left edge (p_l) is greater than on the pressure on the right edge (p_r), i.e., if $\frac{\partial p}{\partial x} < 0$, then flow moves left to right. If the pressure gradient is reversed $\left(\frac{\partial p}{\partial x} > 0\right)$ and the pressure on the right (p_r) is greater than the pressure on the left (p_l), then the fluid flows from right to left.

to be p_l and the pressure on the right edge to be p_r. If p_l is greater than p_r (i.e., if $\frac{\partial p}{\partial x} < 0$), then there is more of an overall "push" on the fluid element to the right (since p_l is pointed to the right direction). Similarly, if the pressure on the right side of the fluid element was greater than the left side (i.e., if $p_r > p_l$), then we would expect a general push on the fluid element to the left. We can see this more clearly by looking at the pressure gradient term and approximating it as a slope:

$$-\frac{\partial p}{\partial x} \approx -\frac{p_r - p_l}{\Delta x} = \frac{p_l - p_r}{\Delta x}$$

where Δx is the length of the fluid element in the x-direction. Notice that if, for example, $p_l > p_r$, then the force due to the pressure will be in the positive x-direction, which is what we expect. Hence the reason the negative sign was included in the pressure gradient term. The next thing to notice is if there is a significant difference between the pressures on the two sides, then there will be a bigger force difference on the fluid element, and hence a larger push on the fluid element. This is also what we would expect. Now, what about the fact that we have a Δx on the denominator? What does that mean? This implies that the larger the fluid element is in the x-direction, the less impact the pressure change will have on the fluid element. In other words, the most important thing to consider when thinking about how pressure affects the movement of the flow, is the larger the pressure change in a given distance (i.e., the larger the pressure gradient), the more significant the impact will be in terms of moving the fluid. In addition, the flow will tend to move from high to low pressure.

Incidentally, you should also note that we would have gotten the same result as Equation 3.31 had we started with the area integral version of the force summation, Equation 3.8, i.e. (assuming only pressure force):

$$\Sigma \vec{F} = \oiint_{A(t)} \overset{\Rightarrow}{T} \cdot \vec{n} dA = \oiint_{A(t)} -p \overset{\Rightarrow}{I} \cdot \vec{n} dA$$

If we use the normals of $\vec{n}_l = -\hat{i}$ and $\vec{n}_r = \hat{i}$, as shown in Figure 3.8, then the above equation expanded out in Cartesian coordinates can be approximated as:

$$\Sigma \vec{F} = \oiint_{A(t)} -p \overset{\Rightarrow}{I} \cdot \vec{n} dA$$

$$\approx \begin{pmatrix} -p_l & 0 & 0 \\ 0 & -p_l & 0 \\ 0 & 0 & -p_l \end{pmatrix} \begin{pmatrix} -1 \\ 0 \\ 0 \end{pmatrix} \underbrace{\Delta y \Delta z}_{\text{area of left face}} + \begin{pmatrix} -p_r & 0 & 0 \\ 0 & -p_r & 0 \\ 0 & 0 & -p_r \end{pmatrix} \begin{pmatrix} 1 \\ 0 \\ 0 \end{pmatrix} \underbrace{\Delta y \Delta z}_{\text{area of right face}}$$

$$\approx \begin{pmatrix} p_l \\ 0 \\ 0 \end{pmatrix} \Delta y \Delta z + \begin{pmatrix} -p_r \\ 0 \\ 0 \end{pmatrix} \Delta y \Delta z$$

where the "*l*" and the "*r*" subscript denote the pressure values at the left and right face, respectively. Dividing out the above equation by the volume of the Cartesian fluid element, i.e., by $\Delta x \Delta y \Delta z$, gives us the force per volume (at a given instant of time) to be:

$$\frac{\Sigma \vec{F}}{\Delta x \Delta y \Delta z} \approx \begin{pmatrix} \frac{p_l - p_r}{\Delta x} \\ 0 \\ 0 \end{pmatrix} \rightarrow \underbrace{\frac{p_l - p_r}{\Delta x}}_{-\frac{\partial p}{\partial x}}$$

This leaves us with what we would expect for the pressure gradient in the *x*-direction.

Case of Zero Viscosity: Euler's Equation

If we assume an inviscid flow, i.e. $\mu = 0$, then the Navier–Stokes equations, Equation 3.29, simplifies to an equation called Euler's equation:

$$\rho \left(\frac{\partial \vec{V}}{\partial t} + \vec{V} \cdot \vec{\nabla} \vec{V} \right) = -\vec{\nabla} p + \rho \vec{g} \tag{3.32}$$

We can see that the only forces acting on an inviscid flow are pressure forces and body forces (in this case, gravitational body forces). If we just consider the

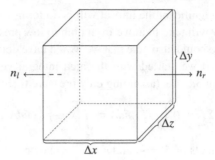

Figure 3.8 Illustration of a Cartesian element with normals of the left face, i.e. $\vec{n}_l = -\hat{i}$ and right face, i.e., $\vec{n}_r = \hat{i}$. The area of the left face and right face are both $\Delta y \Delta z$.

x-component and ignore gravity, we can write Euler's equation (in Lagrangian form) as:

$$\rho \frac{Du}{Dt} = -\frac{\partial p}{\partial x}$$

The left side is mass per volume (ρ) times acceleration $\left(\frac{Du}{Dt}\right)$ and the right side is the pressure acting on the fluid element. Thus, if the pressure gradient is negative $\left(\text{i.e., if pressure decreases with increasing } x \text{ or } \frac{\partial p}{\partial x} < 0\right)$, then the velocity of the fluid element should accelerate to the right because $\frac{Du}{Dt} > 0$, and visa versa. This should make sense because the pressure is larger on the left side of the fluid element than on the right side, thus there would be an imbalance pushing the fluid element to the right.

Static Case and Hydrostatic Pressure

Before moving on to the other terms in Navier–Stokes equations, it is a useful exercise to consider a situation in which there is no fluid motion. In the case of zero velocity, the Navier–Stokes equations would simplify to the following:

$$\left(\cancel{\frac{\partial \vec{V}}{\partial t}} + \cancel{\vec{V} \cdot \vec{\nabla}\vec{V}}\right) = -\vec{\nabla}p - \cancel{\vec{\nabla}\left(\frac{2}{3}\mu\left(\vec{\nabla}\cdot\vec{V}\right)\right)} + \cancel{\vec{\nabla}\cdot\left(\mu\left(\vec{\nabla}\vec{V} + \left(\vec{\nabla}\vec{V}\right)^T\right)\right)} + \rho\vec{g}$$

$$0 = -\vec{\nabla}p + \rho\vec{g}$$

If there was no body force, only a pressure term would describe the system. Leaving us with the following in Cartesian coordinates:

$$\underbrace{\begin{pmatrix} \frac{\partial p}{\partial x} \\ \frac{\partial p}{\partial y} \\ \frac{\partial p}{\partial z} \end{pmatrix}}_{\vec{\nabla}p} = 0 \qquad (3.33)$$

In this particular case, since the derivatives of pressure in all three directions are zero, the pressure value would be a constant value. The source of this pressure would stem from the molecular collisions within the fluid. In particular, the pressure will be related to how closely packed the molecules are in the fluid (which manifests as a density) as well as how fast the molecules in the fluid are moving (which manifests as a temperature). Thus, the pressure in this instance would be related to density and temperature, which indicates a thermodynamic origin. Therefore, the pressure in this case is considered a thermodynamic pressure.

Now, what if we included a gravitational body force along with the pressure gradient? For such a case, we have:

$$\underbrace{-\begin{pmatrix} \frac{\partial p}{\partial x} \\ \frac{\partial p}{\partial y} \\ \frac{\partial p}{\partial z} \end{pmatrix}}_{-\vec{\nabla}p} + \underbrace{\rho\begin{pmatrix} g_x \\ g_y \\ g_z \end{pmatrix}}_{\rho\vec{g}} = 0 \qquad (3.34)$$

where g_x, g_y, and g_z are the x-, y-, and z-coordinates of the gravitational acceleration vector.

Suppose we have a situation where the gravity is only in the y-direction, as seen in Figure 3.9. Then Equation 3.34 becomes:

$$-\begin{pmatrix} \frac{\partial p}{\partial x} \\ \frac{\partial p}{\partial y} \\ \frac{\partial p}{\partial z} \end{pmatrix} + \rho\begin{pmatrix} 0 \\ g_y \\ 0 \end{pmatrix} = 0 \qquad (3.35)$$

We have three separate equations, one for each direction. Integrating the x-direction leads to the possibility that pressure can only be a function of y and z, but not x. Integrating the z-direction leads to the possibility that pressure can only be a function of y and x, but not z. Integrating the y-direction leads to (if

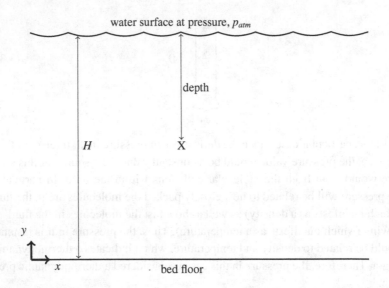

Figure 3.9 Schematic of hydrostatic pressure. The pressure at 'X' is given by:
$p = \rho|g_y|depth + p_{atm}$.

we assume that the gravitational acceleration is a constant in the y-direction):

$$p = \rho g_y y + C \qquad (3.36)$$

where C is a constant of integration. The determination of the constant of integration requires us to apply what is called a boundary condition. That is, a value for the pressure at a known value for y. In order to apply a boundary condition, an axis location, or datum location, needs to be defined. We can set our Cartesian axis anywhere we like but a good place might be at the bottom of our fluid (the bed floor) as shown in Figure 3.9. For example, maybe the bottom of our fluid is the bottom of a lake. Now suppose we know the value of pressure at the top of our fluid and that the top of the fluid is located at $y = H$. The value of pressure at the top could be atmospheric pressure, p_{atm}. Then, by applying the following boundary condition to Equation 3.36:

$$\text{at } y = H, \quad p = p_{atm}$$

we get:

$$p_{atm} = \rho g_y H + C \quad \Longrightarrow \quad C = p_{atm} - \rho g_y H$$

Plugging the $C = p_{atm} - \rho g_y H$ into Equation 3.36 we get:

$$p = \rho g_y y + p_{atm} - \rho g_y H$$
$$= \rho g_y (y - H) + p_{atm}$$

Considering that g_y is a negative value since it is directed downward, we can write the magnitude of g_y as $|g_y|$ and flip the sign of the $y - H$ to get:

$$p = \rho |g_y| (H - y) + p_{atm}$$

Recognizing that the $H - y$ is describing the depth of the fluid from the top surface (i.e., how deep you are), we can write pressure in terms of depth to get:

$$p = \rho |g_y| depth + p_{atm} \qquad (3.37)$$

The pressure given by the expression in Equation 3.37 is called the **hydrostatic pressure**. This equation implies that the further you dive down into the fluid (water most likely in this case), the greater the pressure will be.

Is this pressure a thermodynamic pressure? You will find that this pressure is not really a thermodynamic pressure as it is not inherently dependent on a thermodynamic equation of state but instead on the gravitational force. Pressure is a somewhat unique animal in that it is not always considered a thermodynamic variable as one might expect but can vacillate between a thermodynamic and mechanical meaning. You may recall from Chapter 1 that density had a similar identity crisis. As you will see, pressure in an incompressible flow is also not considered a thermodynamic variable.

3.4.2 Volume Change Term

The next term in the Navier–Stokes equations stems from the force associated with the volume change of the fluid. Note that this term will vanish for incompressible flows since $\vec{\nabla} \cdot \vec{V} = 0$ for incompressible flows. This term is boxed below:

$$\rho \left(\frac{\partial \vec{V}}{\partial t} + \vec{V} \cdot \vec{\nabla} \vec{V} \right) = -\vec{\nabla} p \boxed{-\vec{\nabla} \left(\frac{2}{3} \mu \left(\vec{\nabla} \cdot \vec{V} \right) \right)} + \vec{\nabla} \cdot \left(\mu \left(\vec{\nabla} \vec{V} + \left(\vec{\nabla} \vec{V} \right)^\dagger \right) \right) + \rho \vec{g}$$

This term, like the pressure gradient term, comes directly from applying the divergence of the Newtonian stress tensor, written here:

$$\vec{\nabla} \cdot \vec{T} = \vec{\nabla} \cdot \left(-p \vec{I} - \frac{2}{3} \mu \left(\vec{\nabla} \cdot \vec{V} \right) \vec{I} + \mu \left(\vec{\nabla} \vec{V} + \left(\vec{\nabla} \vec{V} \right)^\dagger \right) \right)$$
$$= \vec{\nabla} \cdot \left(-p \vec{I} \right) + \boxed{\vec{\nabla} \cdot \left(-\frac{2}{3} \mu \left(\vec{\nabla} \cdot \vec{V} \right) \vec{I} \right)} + \vec{\nabla} \cdot \left(\mu \left(\vec{\nabla} \vec{V} + \left(\vec{\nabla} \vec{V} \right)^\dagger \right) \right)$$

Taking the divergence of this term in Cartesian coordinates leads to:

$$
\begin{pmatrix} \frac{\partial}{\partial x} & \frac{\partial}{\partial y} & \frac{\partial}{\partial z} \end{pmatrix}
\begin{pmatrix}
-\frac{2}{3}\mu\left(\vec{\nabla}\cdot\vec{V}\right) & 0 & 0 \\
0 & -\frac{2}{3}\mu\left(\vec{\nabla}\cdot\vec{V}\right) & 0 \\
0 & 0 & -\frac{2}{3}\mu\left(\vec{\nabla}\cdot\vec{V}\right)
\end{pmatrix}
$$

$$
= -\underbrace{\left(\frac{\partial\left(\frac{2}{3}\mu(\vec{\nabla}\cdot\vec{V})\right)}{\partial x} \quad \frac{\partial\left(\frac{2}{3}\mu(\vec{\nabla}\cdot\vec{V})\right)}{\partial y} \quad \frac{\partial\left(\frac{2}{3}\mu(\vec{\nabla}\cdot\vec{V})\right)}{\partial z} \right)}_{-\vec{\nabla}\left(\frac{2}{3}\mu(\vec{\nabla}\cdot\vec{V})\right)}
\tag{3.38}
$$

Notice how we did the exact same thing in Equation 3.38 as we did for pressure in Equation 3.30. That is, applying the divergence to this term resulted in a gradient.

Like before with the pressure gradient, we are going to write the second term as a column vector:

$$
\vec{\nabla}\cdot\left(-\frac{2}{3}\mu\left(\vec{\nabla}\cdot\vec{V}\right)\vec{I}\right) = -\vec{\nabla}\left(\frac{2}{3}\mu\left(\vec{\nabla}\cdot\vec{V}\right)\right) = -
\begin{pmatrix}
\frac{\partial\left(\frac{2}{3}\mu(\vec{\nabla}\cdot\vec{V})\right)}{\partial x} \\[2mm]
\frac{\partial\left(\frac{2}{3}\mu(\vec{\nabla}\cdot\vec{V})\right)}{\partial y} \\[2mm]
\frac{\partial\left(\frac{2}{3}\mu(\vec{\nabla}\cdot\vec{V})\right)}{\partial z}
\end{pmatrix}
\tag{3.39}
$$

In some ways, we can think of this term like we do the pressure gradient term, except instead of pressure, we are using the divergence of velocity. Recall that the pressure gradient term suggested that the flow "moves" in the direction of high to low pressure. Likewise, flow would have a tendency to "move" in a direction of high $\frac{2}{3}\mu\left(\vec{\nabla}\cdot\vec{V}\right)$ to low $\frac{2}{3}\mu\left(\vec{\nabla}\cdot\vec{V}\right)$. To give an idea of why this might be the case, recall that the divergence of velocity for a moving element is related to the change in volume of the element, i.e., $\frac{D\mathcal{V}}{Dt} = \iiint_{V(t)} \vec{\nabla}\cdot\vec{V}dV$. If the volume of a fluid element increases (and hence the divergence of velocity increases), there will be a tendency to "push" a fluid element with a smaller change in volume. To help provide a picture (albeit not a perfect picture), such a scenario is illustrated in Figure 3.10. Figure 3.10 shows a number of fluid elements initially all of the same size. The fluid elements in the middle two columns increase in size (thereby having a high $\vec{\nabla}\cdot\vec{V}$ value). This increase in size effectively takes up more space than the fluid elements that did not increase in size (and hence have a small, in this case zero, $\vec{\nabla}\cdot\vec{V}$ value). This will manifest itself as force essentially pushing the smaller fluid elements away causing a flow originating in the middle columns.

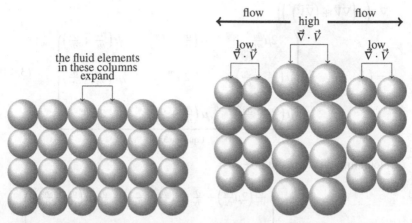

(a) Initial starting configuration of the fluid elements. The middle columns begin to experience an increase in volume, leading to the scenario in Figure 3.10b.

(b) Due to the volume change, the increase in volume of the middle fluid elements "displaces" the other surrounding fluid elements.

Figure 3.10 Illustration of the volume change term in the Navier–Stokes equations. The fluid elements in the middle columns begin to expand, thus $\frac{DV}{Dt}$ (and hence $\vec{\nabla} \cdot \vec{V}$) is larger in the middle columns than the surrounding columns. This leads to a displacement (or a flow) of fluid elements getting "pushed" from high $\vec{\nabla} \cdot \vec{V}$ to low $\vec{\nabla} \cdot \vec{V}$.

3.4.3 The Diffusive Transport Term (Friction Force)

The final part of the surface force in the Navier–Stokes equations generally deals with friction forces and is called the diffusive transport term. The combination of this term and the volume change term described in the last section, i.e., $-\frac{2}{3}\left(\mu\left(\vec{\nabla} \cdot \vec{V}\right)\right)$, make up what is called the viscous force. In incompressible flows, since $\vec{\nabla} \cdot \vec{V} = 0$, the volume change term goes away and thus the diffusive transport term will be the same as the viscous force term. The reasons for calling this a diffusion term will become more clear in Chapter 5 when we discuss the energy equation. The diffusive transport term is boxed below:

$$\rho\frac{D\vec{V}}{Dt} = -\vec{\nabla}p - \vec{\nabla}\left(\frac{2}{3}\mu\left(\vec{\nabla} \cdot \vec{V}\right)\right) + \boxed{\vec{\nabla} \cdot \left(\mu\left(\vec{\nabla}\vec{V} + \left(\vec{\nabla}\vec{V}\right)^{\dagger}\right)\right)} + \rho\vec{g}$$

We can expand this term out in Cartesian coordinates in the following manner:

$$\vec{\nabla} \cdot \left(\mu \left(\vec{\nabla}\vec{V} + \left(\vec{\nabla}\vec{V}\right)^{\dagger} \right) \right)$$

$$= \begin{pmatrix} \frac{\partial}{\partial x} & \frac{\partial}{\partial y} & \frac{\partial}{\partial z} \end{pmatrix} \begin{pmatrix} 2\mu\frac{\partial u}{\partial x} & \mu\left(\frac{\partial v}{\partial x} + \frac{\partial u}{\partial y}\right) & \mu\left(\frac{\partial w}{\partial x} + \frac{\partial u}{\partial z}\right) \\ \mu\left(\frac{\partial u}{\partial y} + \frac{\partial v}{\partial x}\right) & 2\mu\frac{\partial v}{\partial y} & \mu\left(\frac{\partial w}{\partial y} + \frac{\partial v}{\partial z}\right) \\ \mu\left(\frac{\partial u}{\partial z} + \frac{\partial w}{\partial x}\right) & \mu\left(\frac{\partial v}{\partial z} + \frac{\partial w}{\partial y}\right) & 2\mu\frac{\partial w}{\partial z} \end{pmatrix} \tag{3.40}$$

Equation 3.21

Calculating this term out gives us:

$$\vec{\nabla} \cdot \left(\mu \left(\vec{\nabla}\vec{V} + \left(\vec{\nabla}\vec{V}\right)^{\dagger} \right) \right) = \begin{pmatrix} \frac{\partial}{\partial x}\left(2\mu\frac{\partial u}{\partial x}\right) + \frac{\partial}{\partial y}\left(\mu\left(\frac{\partial u}{\partial y} + \frac{\partial v}{\partial x}\right)\right) + \frac{\partial}{\partial z}\left(\mu\left(\frac{\partial u}{\partial z} + \frac{\partial w}{\partial x}\right)\right) \\ \frac{\partial}{\partial x}\left(\mu\left(\frac{\partial v}{\partial x} + \frac{\partial u}{\partial y}\right)\right) + \frac{\partial}{\partial y}\left(2\mu\frac{\partial v}{\partial y}\right) + \frac{\partial}{\partial z}\left(\mu\left(\frac{\partial v}{\partial z} + \frac{\partial w}{\partial y}\right)\right) \\ \frac{\partial}{\partial x}\left(\mu\left(\frac{\partial w}{\partial x} + \frac{\partial u}{\partial z}\right)\right) + \frac{\partial}{\partial y}\left(\mu\left(\frac{\partial w}{\partial y} + \frac{\partial v}{\partial z}\right)\right) + \frac{\partial}{\partial z}\left(2\mu\frac{\partial w}{\partial z}\right) \end{pmatrix}^{\dagger} \tag{3.41}$$

The result is actually a row vector (hence the transpose). However, like what we have done with the gradient of pressure and velocity divergence (i.e., the first and second terms of the stress tensor), when we write this term out in Cartesian coordinates, we will write it out as a column (as opposed to row) vector. Thus, from here on out, we will drop the transpose. We can write out explicitly this force (which we will just call diffusion force) in x-, y-, and z-directions:

$$\begin{pmatrix} \text{diffusion force in x} \\ \text{diffusion force in y} \\ \text{diffusion force in z} \end{pmatrix} = \begin{pmatrix} \frac{\partial}{\partial x}\left(2\mu\frac{\partial u}{\partial x}\right) + \frac{\partial}{\partial y}\left(\mu\left(\frac{\partial u}{\partial y} + \frac{\partial v}{\partial x}\right)\right) + \frac{\partial}{\partial z}\left(\mu\left(\frac{\partial u}{\partial z} + \frac{\partial w}{\partial x}\right)\right) \\ \frac{\partial}{\partial x}\left(\mu\left(\frac{\partial v}{\partial x} + \frac{\partial u}{\partial y}\right)\right) + \frac{\partial}{\partial y}\left(2\mu\frac{\partial v}{\partial y}\right) + \frac{\partial}{\partial z}\left(\mu\left(\frac{\partial v}{\partial z} + \frac{\partial w}{\partial y}\right)\right) \\ \frac{\partial}{\partial x}\left(\mu\left(\frac{\partial w}{\partial x} + \frac{\partial u}{\partial z}\right)\right) + \frac{\partial}{\partial y}\left(\mu\left(\frac{\partial w}{\partial y} + \frac{\partial v}{\partial z}\right)\right) + \frac{\partial}{\partial z}\left(2\mu\frac{\partial w}{\partial z}\right) \end{pmatrix} \tag{3.42}$$

To study this force, we are going to focus our attention on the x-direction, as the extension to the y- and z-directions will use a similar thought process. In addition, to simplify things a touch further, we are going to assume a two-dimensional scenario. That is, we are going to ignore any derivatives in the z-direction. Thus, we are focusing on the following:

$$\text{diffusion force in x} = \underbrace{\frac{\partial}{\partial x}\left(2\mu\frac{\partial u}{\partial x}\right)}_{\text{from } \tau_{xx}} + \underbrace{\frac{\partial}{\partial y}\left(\mu\left(\frac{\partial u}{\partial y} + \frac{\partial v}{\partial x}\right)\right)}_{\text{from } \tau_{yx}} \tag{3.43}$$

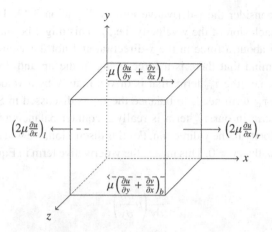

Figure 3.11 Illustration of the diffusive transport term in the x-direction (we are going to ignore the front and back faces).

In Equation 3.43, we are explicitly referencing the different terms back to their original stress tensor origin. Recall from Equation 3.13 that the derivative in x is taken on the τ_{xx} term and the derivative in y is taken on the τ_{yx} term when looking at the force in the x-direction. We can draw a Cartesian stress element to get a visual of the diffusive stresses acting on the element. Such an element is given in Figure 3.11.

The first part of the diffusion force to consider is the derivative of τ_{xx} with respect to x. In Figure 3.11, we marked the values of $2\mu \frac{\partial u}{\partial x}$ on the left and right face with an l and r subscript, respectively. The derivative of this term can then be approximated as:

$$\frac{\partial}{\partial x}\left(2\mu\frac{\partial u}{\partial x}\right) \approx \frac{1}{\Delta x}\left(\left(2\mu\frac{\partial u}{\partial x}\right)_r - \left(2\mu\frac{\partial u}{\partial x}\right)_l\right)$$

where Δx is the distance between the left and the right face. Thus, if the value of $2\mu\frac{\partial u}{\partial x}$ is larger on the right than on the left, then the overall force on the fluid element is to the right (or in the positive x-direction), and visa-versa. The key to visualizing this is by recognizing that a positive stress value on the left side points in the negative x-direction while a positive stress value on the right side points in the positive x-direction, as shown in Figure 3.11. This hearkens back to how we defined stresses in Figure 3.2. This is the opposite of the pressure and volume change terms. Recall, as an example, that a positive pressure value is compressive and the pressure gradient has a minus sign in front.

Let's now consider the y-derivative term in Equation 3.43. The first thing to note is the inclusion of the y-velocity, i.e., v. This might be surprising since we are talking about a force in the x-direction and not the y-direction. However, keep in mind that there is a force going in the up and down direction on the x-faces, i.e., the τ_{xy} term, that is driven mainly by a velocity in the y-direction. The τ_{xy} term needs to balance the τ_{yx} as discussed in Section 3.1.1. Thus, the $\frac{\partial v}{\partial x}$ term in our τ_{yx} term is really a counter-balancing force in case there was a velocity in the y-direction. For this discussion, we will assume the y-velocity is zero, thus $v = 0$. This means the y-derivative term in Equation 3.43 is now just:

$$\frac{\partial}{\partial y}\left(\mu\frac{\partial u}{\partial y}\right)$$

We can approximate this term via the following:

$$\frac{\partial}{\partial y}\left(\mu\frac{\partial u}{\partial y}\right) \approx \frac{1}{\Delta y}\left(\left(\mu\frac{\partial u}{\partial y}\right)_t - \left(\mu\frac{\partial u}{\partial y}\right)_b\right)$$

where the subscripts "t" and "b" mean top and bottom, respectively, and the Δy is the distance between the top and bottom surfaces. We discussed the $\mu\frac{\partial u}{\partial y}$ before when we discussed the τ_{yx} term in Section 3.3. We now see that the difference in this term between the top and the bottom surface of our Cartesian element will cause a force on a fluid element in the x-direction. The force will be in the positive x-direction if $\mu\frac{\partial u}{\partial y}$ is larger at the top than at the bottom, and visa-versa. Another way of saying it is if $\mu\frac{\partial u}{\partial y}$ increases with increasing y, then the overall contribution to the force will be to the right. If $\mu\frac{\partial u}{\partial y}$ decreases with increasing y, then the overall contribution to the force will be to the left.

3.5 Incompressible Flow

We now turn our attention to incompressible flows. A great many situations often arise where the flow can be considered incompressible. Many situations involving liquids (as opposed to gases) can be modeled as incompressible flow. Since an incompressible flow has a divergence-free velocity field (i.e. $\vec{\nabla}\cdot\vec{V} = 0$), the Newtonian stress tensor in Equation 3.26 becomes the **Newtonian fluid stress tensor for an incompressible flow**:

$$\boxed{\vec{\vec{T}}_{incompressible} = -p\vec{\vec{I}} + \mu\left(\vec{\nabla}\vec{V} + \left(\vec{\nabla}\vec{V}\right)^\dagger\right)} \tag{3.44}$$

The divergence of the incompressible stress tensor looks like the following:

$$\vec{\nabla} \cdot \vec{\vec{T}}_{incompressible} = \underbrace{\vec{\nabla} \cdot \left(-p\vec{\vec{I}}\right)}_{\substack{-\vec{\nabla}p \\ \text{(like before)}}} + \vec{\nabla} \cdot \left(\mu\left(\vec{\nabla}\vec{V} + \left(\vec{\nabla}\vec{V}\right)^{\dagger}\right)\right) \tag{3.45}$$

$$= -\vec{\nabla}p + \vec{\nabla} \cdot \left(\mu\left(\vec{\nabla}\vec{V} + \left(\vec{\nabla}\vec{V}\right)^{\dagger}\right)\right)$$

The divergence of the pressure term is easy and is the same as we did before, leaving us with the negative of the gradient of pressure (i.e., $-\vec{\nabla}p$). The divergence of the shearing term (i.e., the $\vec{\nabla} \cdot \left(\mu\left(\vec{\nabla}\vec{V} + \left(\vec{\nabla}\vec{V}\right)^{\dagger}\right)\right)$ term) can also be simplified if we make the additional assumption of a constant dynamic viscosity (which is not an unrealistic assumption for a lot of incompressible flow situations). To illustrate the simplifications we can make, it will be easier to break out and look at the terms in Cartesian coordinates. This will take a decent amount of arithmetic, but it will be a useful exercise. Recall from the previous section that the divergence of the shearing term can be written in Cartesian coordinates as (in column vector form):

$$\vec{\nabla} \cdot \left(\mu\left(\vec{\nabla}\vec{V} + \vec{\nabla}\vec{V}^{\dagger}\right)\right) = \begin{pmatrix} \frac{\partial}{\partial x}\left(2\mu\frac{\partial u}{\partial x}\right) + \frac{\partial}{\partial y}\left(\mu\left(\frac{\partial u}{\partial y} + \frac{\partial v}{\partial x}\right)\right) + \frac{\partial}{\partial z}\left(\mu\left(\frac{\partial u}{\partial z} + \frac{\partial w}{\partial x}\right)\right) \\ \frac{\partial}{\partial x}\left(\mu\left(\frac{\partial v}{\partial x} + \frac{\partial u}{\partial y}\right)\right) + \frac{\partial}{\partial y}\left(2\mu\frac{\partial v}{\partial y}\right) + \frac{\partial}{\partial z}\left(\mu\left(\frac{\partial v}{\partial z} + \frac{\partial w}{\partial y}\right)\right) \\ \frac{\partial}{\partial x}\left(\mu\left(\frac{\partial w}{\partial x} + \frac{\partial u}{\partial z}\right)\right) + \frac{\partial}{\partial y}\left(\mu\left(\frac{\partial w}{\partial y} + \frac{\partial v}{\partial z}\right)\right) + \frac{\partial}{\partial z}\left(2\mu\frac{\partial w}{\partial z}\right) \end{pmatrix}$$

Assuming a constant dynamic viscosity allows us to pull out the dynamic viscosity completely from the stress tensor "matrix" as well as "distribute" the derivatives into the parentheses:

$$\vec{\nabla} \cdot \left(\mu\left(\vec{\nabla}\vec{V} + \vec{\nabla}\vec{V}^{\dagger}\right)\right) = \mu \begin{pmatrix} 2\frac{\partial}{\partial x}\left(\frac{\partial u}{\partial x}\right) + \left(\frac{\partial}{\partial y}\left(\frac{\partial u}{\partial y}\right) + \frac{\partial}{\partial y}\left(\frac{\partial v}{\partial x}\right)\right) + \left(\frac{\partial}{\partial z}\left(\frac{\partial u}{\partial z}\right) + \frac{\partial}{\partial z}\left(\frac{\partial w}{\partial x}\right)\right) \\ \left(\frac{\partial}{\partial x}\left(\frac{\partial v}{\partial x}\right) + \frac{\partial}{\partial x}\left(\frac{\partial u}{\partial y}\right)\right) + 2\frac{\partial}{\partial y}\left(\frac{\partial v}{\partial y}\right) + \left(\frac{\partial}{\partial z}\left(\frac{\partial v}{\partial z}\right) + \frac{\partial}{\partial z}\left(\frac{\partial w}{\partial y}\right)\right) \\ \left(\frac{\partial}{\partial x}\left(\frac{\partial w}{\partial x}\right) + \frac{\partial}{\partial x}\left(\frac{\partial u}{\partial z}\right)\right) + \left(\frac{\partial}{\partial y}\left(\frac{\partial w}{\partial y}\right) + \frac{\partial}{\partial y}\left(\frac{\partial v}{\partial z}\right)\right) + 2\frac{\partial}{\partial z}\left(\frac{\partial w}{\partial z}\right) \end{pmatrix}$$

Next we can split apart the $2\frac{\partial}{\partial x}\left(\frac{\partial u}{\partial x}\right)$ into $\frac{\partial}{\partial x}\left(\frac{\partial u}{\partial x}\right) + \frac{\partial}{\partial x}\left(\frac{\partial u}{\partial x}\right)$ and the same for the $2\frac{\partial}{\partial y}\left(\frac{\partial v}{\partial y}\right)$ and $2\frac{\partial}{\partial z}\left(\frac{\partial w}{\partial z}\right)$ terms to get:

$$\vec{\nabla} \cdot \left(\mu \left(\vec{\nabla} \vec{V} + \vec{\nabla} \vec{V}^\dagger \right) \right)$$

$$= \mu \begin{pmatrix} \left(\frac{\partial}{\partial x} \left(\frac{\partial u}{\partial x} \right) + \frac{\partial}{\partial x} \left(\frac{\partial u}{\partial x} \right) \right) + \left(\frac{\partial}{\partial y} \left(\frac{\partial u}{\partial y} \right) + \frac{\partial}{\partial y} \left(\frac{\partial v}{\partial x} \right) \right) + \left(\frac{\partial}{\partial z} \left(\frac{\partial u}{\partial z} \right) + \frac{\partial}{\partial z} \left(\frac{\partial w}{\partial x} \right) \right) \\[2mm] \left(\frac{\partial}{\partial x} \left(\frac{\partial v}{\partial x} \right) + \frac{\partial}{\partial x} \left(\frac{\partial u}{\partial y} \right) \right) + \frac{\partial}{\partial y} \left(\frac{\partial v}{\partial y} \right) + \frac{\partial}{\partial y} \left(\frac{\partial v}{\partial y} \right) + \left(\frac{\partial}{\partial z} \left(\frac{\partial v}{\partial z} \right) + \frac{\partial}{\partial z} \left(\frac{\partial w}{\partial y} \right) \right) \\[2mm] \left(\frac{\partial}{\partial x} \left(\frac{\partial w}{\partial x} \right) + \frac{\partial}{\partial x} \left(\frac{\partial u}{\partial z} \right) \right) + \left(\frac{\partial}{\partial y} \left(\frac{\partial w}{\partial y} \right) + \frac{\partial}{\partial y} \left(\frac{\partial v}{\partial z} \right) \right) + \frac{\partial}{\partial z} \left(\frac{\partial w}{\partial z} \right) + \frac{\partial}{\partial z} \left(\frac{\partial w}{\partial z} \right) \end{pmatrix}$$

We can reorder some of the terms in each row to get:

$$\vec{\nabla} \cdot \left(\mu \left(\vec{\nabla} \vec{V} + \vec{\nabla} \vec{V}^\dagger \right) \right)$$

$$= \mu \begin{pmatrix} \frac{\partial}{\partial x} \left(\frac{\partial u}{\partial x} \right) + \frac{\partial}{\partial y} \left(\frac{\partial u}{\partial y} \right) + \frac{\partial}{\partial z} \left(\frac{\partial u}{\partial z} \right) + \frac{\partial}{\partial x} \left(\frac{\partial u}{\partial x} \right) + \frac{\partial}{\partial y} \left(\frac{\partial v}{\partial x} \right) + \frac{\partial}{\partial z} \left(\frac{\partial w}{\partial x} \right) \\[2mm] \frac{\partial}{\partial x} \left(\frac{\partial v}{\partial x} \right) + \frac{\partial}{\partial y} \left(\frac{\partial v}{\partial y} \right) + \frac{\partial}{\partial z} \left(\frac{\partial v}{\partial z} \right) + \frac{\partial}{\partial x} \left(\frac{\partial u}{\partial y} \right) + \frac{\partial}{\partial y} \left(\frac{\partial v}{\partial y} \right) + \frac{\partial}{\partial z} \left(\frac{\partial w}{\partial y} \right) \\[2mm] \frac{\partial}{\partial x} \left(\frac{\partial w}{\partial x} \right) + \frac{\partial}{\partial y} \left(\frac{\partial w}{\partial y} \right) + \frac{\partial}{\partial z} \left(\frac{\partial w}{\partial z} \right) + \frac{\partial}{\partial x} \left(\frac{\partial u}{\partial z} \right) + \frac{\partial}{\partial y} \left(\frac{\partial v}{\partial z} \right) + \frac{\partial}{\partial z} \left(\frac{\partial w}{\partial z} \right) \end{pmatrix}$$

Finally, we can rearrange the last three terms of each row by "factoring out" $\frac{\partial}{\partial x}$ in the first row, $\frac{\partial}{\partial y}$ in the second row, and $\frac{\partial}{\partial z}$ in the third row. In addition, for the first three terms of each row, we can now make them second derivatives. Thus:

$$\vec{\nabla} \cdot \left(\mu \left(\vec{\nabla} \vec{V} + \vec{\nabla} \vec{V}^\dagger \right) \right) = \mu \begin{pmatrix} \left(\frac{\partial^2 u}{\partial x^2} + \frac{\partial^2 u}{\partial y^2} + \frac{\partial^2 u}{\partial z^2} \right) + \frac{\partial}{\partial x} \left(\frac{\partial u}{\partial x} + \frac{\partial v}{\partial y} + \frac{\partial w}{\partial z} \right) \\[2mm] \left(\frac{\partial^2 v}{\partial x^2} + \frac{\partial^2 v}{\partial y^2} + \frac{\partial^2 v}{\partial z^2} \right) + \frac{\partial}{\partial y} \left(\frac{\partial u}{\partial x} + \frac{\partial v}{\partial y} + \frac{\partial w}{\partial z} \right) \\[2mm] \left(\frac{\partial^2 w}{\partial x^2} + \frac{\partial^2 w}{\partial y^2} + \frac{\partial^2 w}{\partial z^2} \right) + \frac{\partial}{\partial z} \left(\frac{\partial u}{\partial x} + \frac{\partial v}{\partial y} + \frac{\partial w}{\partial z} \right) \end{pmatrix}$$

The first three terms of each row (e.g., the $\mu \left(\frac{\partial^2 u}{\partial x^2} + \frac{\partial^2 u}{\partial y^2} + \frac{\partial^2 u}{\partial z^2} \right)$ term in the first row) can be written using the Laplacian operator. The Laplacian of a generic function, f, is usually written in del form as:

$$\nabla^2 f = \vec{\nabla} \cdot \vec{\nabla} f$$

The Laplacian of the u, v, and w velocities are then written as $\nabla^2 u$, $\nabla^2 v$, and $\nabla^2 w$, respectively. The reason we sometimes write the Laplacian as $\vec{\nabla} \cdot \vec{\nabla}$ is because the "dot" product operation with the del operator leads to the Laplacian operator. Observe:

$$\nabla^2 = \vec{\nabla} \cdot \vec{\nabla} = \begin{pmatrix} \frac{\partial}{\partial x} & \frac{\partial}{\partial y} & \frac{\partial}{\partial z} \end{pmatrix} \begin{pmatrix} \frac{\partial}{\partial x} \\ \frac{\partial}{\partial y} \\ \frac{\partial}{\partial z} \end{pmatrix} = \underbrace{\frac{\partial^2}{\partial x^2} + \frac{\partial^2}{\partial y^2} + \frac{\partial^2}{\partial z^2}}_{\substack{\text{Laplacian operator in} \\ \text{Cartesian coordinates}}} \tag{3.46}$$

We can now focus on the last three terms in each row (e.g., $\frac{\partial}{\partial x}\left(\frac{\partial u}{\partial x} + \frac{\partial v}{\partial y} + \frac{\partial w}{\partial z}\right)$ from the first row). Notice that, for each direction, there is a common term, the $\frac{\partial u}{\partial x} + \frac{\partial v}{\partial y} + \frac{\partial w}{\partial z}$ term. This is the divergence of velocity, $\vec{\nabla} \cdot \vec{V}$. Using this fact gives us:

$$\vec{\nabla} \cdot \left(\mu\left(\vec{\nabla}\vec{V} + \vec{\nabla}\vec{V}^\dagger\right)\right) = \mu \begin{pmatrix} \nabla^2 u + \frac{\partial}{\partial x}\left(\vec{\nabla} \cdot \vec{V}\right) \\ \nabla^2 v + \frac{\partial}{\partial y}\left(\vec{\nabla} \cdot \vec{V}\right) \\ \nabla^2 w + \frac{\partial}{\partial z}\left(\vec{\nabla} \cdot \vec{V}\right) \end{pmatrix}$$

The Laplacian term in each velocity direction is nothing but the Laplacian of the velocity vector, i.e.:

$$\mu \begin{pmatrix} \nabla^2 u \\ \nabla^2 v \\ \nabla^2 w \end{pmatrix} = \mu\vec{\nabla}^2 \underbrace{\begin{pmatrix} u \\ v \\ w \end{pmatrix}}_{\vec{V}} = \mu\vec{\nabla}^2\vec{V} \tag{3.47}$$

In addition, notice that the derivatives of the divergence term can be written as a gradient, i.e.:

$$\mu \begin{pmatrix} \frac{\partial}{\partial x}\left(\vec{\nabla} \cdot \vec{V}\right) \\ \frac{\partial}{\partial y}\left(\vec{\nabla} \cdot \vec{V}\right) \\ \frac{\partial}{\partial z}\left(\vec{\nabla} \cdot \vec{V}\right) \end{pmatrix} = \mu \underbrace{\begin{pmatrix} \frac{\partial}{\partial x} & \frac{\partial}{\partial y} & \frac{\partial}{\partial z} \end{pmatrix}}_{\vec{\nabla}}\vec{\nabla} \cdot \vec{V}$$

Again, we are assuming a transpose of the gradient. We can now write $\vec{\nabla} \cdot \left(\mu\left(\vec{\nabla}\vec{V} + \vec{\nabla}\vec{V}^\dagger\right)\right)$ as the following for a constant dynamic viscosity:

$$\vec{\nabla} \cdot \left(\mu\left(\vec{\nabla}\vec{V} + \vec{\nabla}\vec{V}^\dagger\right)\right) = \mu\nabla^2\vec{V} + \mu\vec{\nabla}\left(\vec{\nabla} \cdot \vec{V}\right) \tag{3.48}$$

For an incompressible flow, the divergence of velocity is zero, thus we are left with:

$$\vec{\nabla} \cdot \left(\mu \left(\vec{\nabla}\vec{V} + \left(\vec{\nabla}\vec{V} \right)^{\dagger} \right) \right) = \mu \nabla^2 \vec{V} \tag{3.49}$$

The divergence of the diffusive transport term simply becomes the Laplacian of the velocity vector for an incompressible flow with a constant dynamic viscosity.

Inserting 3.49 into 3.45 leads to the **divergence of the Newtonian fluid stress tensor for an incompressible flow with constant dynamic viscosity**:

$$\boxed{\vec{\nabla} \cdot \vec{\vec{T}}_{incompressible} = -\vec{\nabla}p + \mu \nabla^2 \vec{V}} \tag{3.50}$$

The **divergence of the Newtonian fluid stress tensor for an incompressible flow with constant dynamic viscosity in Cartesian coordinates** is (taking advantage of Equations 3.31, 3.47, and 3.46):

$$\vec{\nabla} \cdot \vec{\vec{T}}_{incompressible} = -\vec{\nabla}p + \mu \nabla^2 \vec{V} = -\begin{pmatrix} \frac{\partial p}{\partial x} \\ \frac{\partial p}{\partial y} \\ \frac{\partial p}{\partial z} \end{pmatrix} + \mu \begin{pmatrix} \frac{\partial^2 u}{\partial x^2} + \frac{\partial^2 u}{\partial y^2} + \frac{\partial^2 u}{\partial z^2} \\ \frac{\partial^2 v}{\partial x^2} + \frac{\partial^2 v}{\partial y^2} + \frac{\partial^2 v}{\partial z^2} \\ \frac{\partial^2 w}{\partial x^2} + \frac{\partial^2 w}{\partial y^2} + \frac{\partial^2 w}{\partial z^2} \end{pmatrix} \tag{3.51}$$

We can now use Equation 3.50 for our surface force term in Cauchy's first law, i.e., Equation 3.11 or $\rho \frac{D\vec{V}}{Dt} = \vec{\nabla} \cdot \vec{\vec{T}} + \rho\vec{g}$, to get the **incompressible Navier–Stokes equations in Lagrangian form**:

$$\boxed{\rho \frac{D\vec{V}}{Dt} = -\vec{\nabla}p + \mu \nabla^2 \vec{V} + \rho\vec{g}} \tag{3.52}$$

Expanding out the material derivative in Equation 3.52 into an Eulerian description yields the **incompressible Navier–Stokes equations in a non-conservation form**:

$$\boxed{\rho \left(\frac{\partial \vec{V}}{\partial t} + \vec{V} \cdot \vec{\nabla}\vec{V} \right) = -\vec{\nabla}p + \mu \nabla^2 \vec{V} + \rho\vec{g}} \tag{3.53}$$

The above equation is the most cited version of the Navier–Stokes equations. Just as we did with the compressible version of the equations, we should label the terms of the incompressible Navier–Stokes equations:

Incompressible Navier–Stokes Equations (non-conservation form)

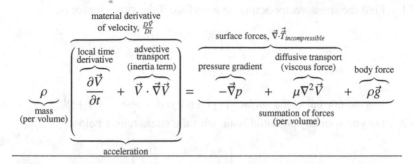

Notice that the incompressible Navier–Stokes equations no longer have a term associated with volume change of a fluid element (i.e., a $\vec{\nabla}\left(\frac{2}{3}\mu\left(\vec{\nabla}\cdot\vec{V}\right)\right)$ term). This should make sense because incompressible flows, by definition, have a divergence-free velocity field. In addition, the diffusive transport term has now been simplified in the incompressible version of the equations to be the Laplacian of the velocity vector. This simplification will come in handy because there will be a similar Laplacian operation in the energy equation that is easier to understand which will be a guide for us in interpreting the diffusive term in the Navier–Stokes equations. Note also that the diffusive transport term, for incompressible flows, is the same as the viscous force term.

There are four unknowns in the incompressible Navier–Stokes equations: the three components of velocity and pressure. Keep in mind that the density is usually considered an input (i.e., a value is provided) to the equations. We have three equations with the incompressible Navier–Stokes equations (an equation for each direction, x-, y-, and z-) and the fourth equation being the incompressible form of the continuity equation (i.e., $\vec{\nabla}\cdot\vec{V} = 0$). Thus, unlike compressible flows, we do not need an equation of state nor do we need an energy equation. This implies that both density and pressure are not to be considered thermodynamic variables since equations involving temperature are not necessary. Pressure is simply a mechanical force acting on the fluid in an incompressible flow.

In addition, notice that there is no separate equation for pressure in an incompressible flow. Pressure only shows up in the pressure gradient term in the incompressible Navier–Stokes equations. Given this fact, only pressure differences matter in an incompressible flow, as opposed to the value of pressure itself. For those of you who are more inclined towards mathematics, pressure is considered a Lagrange multiplier to the incompressible Navier–Stokes equations with the incompressible continuity equation being the constraint.

Problems

3.1   Find the stress vector acting on a surface if the stress tensor is:

$$\vec{\vec{T}} = \begin{pmatrix} 5 & -3 & 10 \\ -3 & 2 & 4 \\ 10 & 4 & 7 \end{pmatrix}$$

and the normal of the surface is given by: $\hat{n} = \frac{1}{\sqrt{2}}\hat{i} + \frac{1}{\sqrt{2}}\hat{j} + 0\hat{k}$.

3.2   Do you spot any potential issue with the stress tensor below?

$$\vec{\vec{T}} = \begin{pmatrix} 5x & 4x & 10z \\ -3y & 2xy & 4x \\ 2z & 4y & 7xyz \end{pmatrix}$$

3.3   A stress tensor (in pascals) of a flow is given by:

$$\vec{\vec{T}} = \begin{pmatrix} 5x & -3y & 10z \\ -3y & 5y & 4x \\ 10z & 4x & 5z \end{pmatrix}$$

What is the acceleration of a fluid particle with density of 1.2kg/m^3 (assuming no body force)?

3.4   If the velocity field of a flow is given by:

$$\vec{V} = A\cos(x)\hat{i} + B\sin(y)\hat{j} + C\tan(z)\hat{k}$$

Determine an expression for the total force (per volume) acting on the fluid at a given (x, y, z) point.

3.5   For the velocity field given in Problem 3.4, if there are no body forces and if the dynamic viscosity is considered a constant, find an expression for the pressure gradient (given $A = 1$, $B = 1$, and $C = 0$). Be sure to check if this flow is incompressible or not. Is this flow at a steady state?

3.6   Determine the Newtonian stress tensor if the pressure is given by $10x$ pascals and a velocity field of $\vec{V} = 2x\sin(2y)\hat{i} + x\cos(2y)\hat{j}$ m/s at location $x = 1$ m, $y = 2$ m. Assume a dynamic viscosity of 10^{-4} Pa·s.

3.7   What is the pressure, in pascals, a person experiences when they have dived 5 meters below the water surface? Assume atmospheric pressure is 10^5 pascals.

3.8   Estimate the shear stress (i.e., viscous force) on a surface defined by $\hat{n} = \hat{i}$ where the velocity distribution is given by:

$$u = \frac{U}{H}y$$

$$v = w = 0$$

where U is a velocity parameter and H is a length scale parameter.

3.9 Show that $\left(\vec{V} \cdot \vec{\nabla}\right) \vec{V} = \vec{V} \cdot \left(\vec{\nabla}\vec{V}\right)$ in two dimensional Cartesian coordinates.

3.10 Please explain, in your own words, why an equation of state is not necessarily needed for an incompressible flow.

4

The Navier–Stokes Equations:
Another Approach

In the previous chapter, we extensively studied the forces acting on a moving fluid element. Using Newton's second law of motion, we were able to obtain the Navier–Stokes equations for both compressible and incompressible flows. There is an alternative approach to developing the Navier–Stokes equations. Namely, we can start from an Eulerian description (as opposed to a Lagrangian description) and develop the integral form and conservation form of the Navier–Stokes equations. We will do that in this chapter. We will then briefly tabulate and discuss the Navier–Stokes equations in its various forms.

We will end this chapter by solving some very simple, yet common, problems involving the incompressible Navier–Stokes equations.

4.1 Eulerian Approach to the Navier–Stokes Equations

In Chapter 1, we stated that the equations of fluid motion can be obtained by applying conservation principles (of mass, momentum, and energy) to a moving fluid passing through a fixed region in space. The Navier–Stokes equations can be obtained by applying the conservation of momentum to a fixed control volume.

Recall the conservation principle stated in Chapter 1:

$$
\begin{pmatrix} \text{the change} \\ \text{of a quantity} \\ \text{in a system} \\ \text{in a given} \\ \text{unit of time} \end{pmatrix} = \begin{pmatrix} \text{time rate of} \\ \text{that quantity} \\ \text{entering the} \\ \text{system} \end{pmatrix} - \begin{pmatrix} \text{time rate of} \\ \text{that quantity} \\ \text{leaving the} \\ \text{system} \end{pmatrix} + \begin{pmatrix} \text{time rate of} \\ \text{a source or} \\ \text{sink inside} \\ \text{the system} \end{pmatrix}
$$

We used this principle to develop the continuity equation in Chapter 1. This approach uses an Eulerian description of the flow since we "fix" our eyes on a

102

particular location (region) in space as opposed to following a fluid element as it travels. We are going to use this approach with momentum as our quantity of interest.

When we applied the conservation balance to a control volume for mass in Chapter 1 we started with a simple expression:

$$\frac{dm_{sys}}{dt} = \dot{m}_{in} - \dot{m}_{out}$$

where m_{sys} was the mass contained in the system (or, in this case, a control volume), $\dot{m}_{in}$ was the mass flow rate into the system and $\dot{m}_{out}$ was the mass flow rate out of the system. Recall that we assumed the control volume had no sources or sinks of mass. However, for the momentum balance, there is now a "source" of momentum. That "source" is the force acting on the fluid. Thus, for the momentum balance, we are going to use a similar starting expression as mass except now we are going to include a summation of force term, that is:

$$\frac{dmom_{sys}}{dt} = \dot{mom}_{in} - \dot{mom}_{out} + \Sigma \vec{F} \tag{4.1}$$

where mom_{sys} is the total momentum of the flow in our control volume, $\dot{mom}_{in}$ is the rate of momentum of the flow entering our control volume, and $\dot{mom}_{out}$ is the rate of momentum of the flow leaving our control volume.

We know that momentum is defined as $m\vec{V}$. However, the momentum of the flow in the control volume can be defined more generally if we integrate over the volume region, that is:

$$mom_{sys} = \iiint_V \rho \vec{V} dV. \tag{4.2}$$

Thus, if we had an arbitrarily shaped volume, the $\rho \vec{V} dV$ would be the momentum of the flow at various infinitesimal volumes within the control volume. Summing up all of the infinitesimal volumes gives us the total momentum of the flow passing through the whole volume.

We can plug Equation 4.2 into Equation 4.1 to get:

$$\frac{d}{dt} \left(\iiint_V \rho \vec{V} dV \right) = \dot{mom}_{in} - \dot{mom}_{out} + \Sigma \vec{F}. \tag{4.3}$$

Now we turn our attention to the rates of momentum in and out of the control volume. Recall in our discussion on mass conservation that we developed the following relationship (i.e., Equation 1.17):

$$\dot{m}_{in} - \dot{m}_{out} = - \oiint_A \rho \vec{V} \cdot \vec{n} dA$$

where $\vec{n}$ was the unit normal on a differential area segment of the control volume and $\vec{V}$ was the velocity of the flow leaving the surface. The quantity $\rho\vec{V}$ was called a mass flux ($\vec{m}''$). Thus, the total mass flow rate passing through the surface of our control volume is related to the area integral of its mass flux via:

$$\dot{m}_{in} - \dot{m}_{out} = - \oiint_A \vec{m}'' \cdot \vec{n}dA.$$

We can do something similar for the rate of momentum passing through the surface of our control volume, such as:

$$\dot{mom}_{in} - \dot{mom}_{out} = - \oiint_A \vec{mom}'' \cdot \vec{n}dA \tag{4.4}$$

where $\vec{mom}''$ is the flux of momentum.

We now have to figure out an expression for $\vec{mom}''$. Considering that the mass flux is density (which is mass per volume) times velocity, maybe we can think of momentum flux in much the same way. That is, as momentum per volume (i.e., density times velocity) times velocity. Thus, the momentum flux for the x-direction would be ρu (momentum per volume) multiplied by $\vec{V}$, or $\rho u\vec{V} = \rho u\left(u\hat{i} + v\hat{j} + w\hat{k}\right)$. Notice that the flux of the x-momentum is a vector quantity. Similarly, the flux of the y-momentum would be: $\rho v\vec{V} = \rho v\left(u\hat{i} + v\hat{j} + w\hat{k}\right)$ and the flux of the z-momentum would look like: $\rho w\vec{V} = \rho w\left(u\hat{i} + v\hat{j} + w\hat{k}\right)$. As you can see, the overall flux of momentum will involve the fluxes of the momentum in the three different component directions (which are vector quantities). Thus, we have multiple directions and their components involved: the components of the momentum and the direction (and components) of the flux. We can encode all of this information in order to establish a general expression for the momentum flux using what is called a tensor product, which is an operation on two vectors (in this case, the velocity vectors), and is usually given by the symbol $\otimes$. In our case, the flux of momentum is defined as:

$$\vec{mom}'' = \rho\vec{V} \otimes \vec{V}. \tag{4.5}$$

Many of you may not have seen the tensor product. The operation of the tensor product is sometimes called taking the outer product. The tensor product of two column vectors, $\vec{A}$ and $\vec{B}$, can be written out as a matrix multiplication of $\vec{A}$ with the transpose of $\vec{B}$ (which is the exact opposite of what would be done for a dot, or inner, product), that is:

$$\vec{A} \otimes \vec{B} = \vec{A}\vec{B}^{\dagger} = \begin{pmatrix} A_x \\ A_y \\ A_z \end{pmatrix} \begin{pmatrix} B_x & B_y & B_z \end{pmatrix}$$

$$= \begin{pmatrix} A_x B_x & A_x B_y & A_x B_z \\ A_y B_x & A_y B_y & A_y B_z \\ A_z B_x & A_z B_y & A_z B_z \end{pmatrix}. \tag{4.6}$$

The outer product operation leads to a second-order tensor and can be represented as a matrix. In our case, the tensor product of the velocity vector with itself yields (in Cartesian coordinates):

$$\vec{V} \otimes \vec{V} = \vec{V}\vec{V}^{\dagger} = \begin{pmatrix} u \\ v \\ w \end{pmatrix} \begin{pmatrix} u & v & w \end{pmatrix}$$

$$= \begin{pmatrix} uu & uv & uw \\ vu & vv & vw \\ wu & wv & ww \end{pmatrix}. \tag{4.7}$$

You can multiply ρ through to get:

$$\rho \vec{V} \otimes \vec{V} = \begin{pmatrix} \rho uu & \rho uv & \rho uw \\ \rho vu & \rho vv & \rho vw \\ \rho wu & \rho wv & \rho ww \end{pmatrix}. \tag{4.8}$$

Notice that Equation 4.8 is a symmetric. Row one is the same as column one, row two is the same as column two, and row three is the same as column three. Notice also, whether you look at the columns or rows, we have captured the flux of the individual momentum components. The first row (column) can be considered the flux of the x-momentum, the second row (column) can be considered the flux of the y-momentum, and the third row (column) is the flux of the z-momentum.

We can plug Equation 4.5 into Equation 4.4 to get an expression for the momentum rate in and out of a control volume:

$$\dot{mom}_{in} - \dot{mom}_{out} = - \oiint_A \rho \vec{V} \otimes \vec{V} \cdot \vec{n} dA. \tag{4.9}$$

The next thing we can do is plug in Equation 4.9 into the expression for momentum conservation, Equation 4.3, to get:

$$\frac{d}{dt}\left(\iiint_V \rho \vec{V} dV \right) = - \oiint_A \rho \vec{V} \otimes \vec{V} \cdot \vec{n} dA + \Sigma \vec{F}. \tag{4.10}$$

We can bring the time derivative into the integral. This time, since the volume is fixed in space and does not change with time, the ordinary derivative just becomes a partial derivative of the integrand via:

$$\iiint_{\mathcal{V}} \frac{\partial\left(\rho\vec{V}\right)}{\partial t}d\mathcal{V} + \oiint_{A}\rho\vec{V}\otimes\vec{V}\cdot\vec{n}dA = \Sigma\vec{F}. \tag{4.11}$$

We can substitute Equation 3.8 for the summation of forces, that is $\Sigma\vec{F} = \oiint_{A}\vec{\vec{T}}\cdot\vec{n}dA + \iiint_{\mathcal{V}}\rho\vec{g}d\mathcal{V}$, into Equation 4.11 to get:

$$\iiint_{\mathcal{V}} \frac{\partial\left(\rho\vec{V}\right)}{\partial t}d\mathcal{V} + \oiint_{A}\rho\vec{V}\otimes\vec{V}\cdot\vec{n}dA = \oiint_{A}\vec{\vec{T}}\cdot\vec{n}dA + \iiint_{\mathcal{V}}\rho\vec{g}d\mathcal{V}. \tag{4.12}$$

Recognize now that we are integrating over a control volume that does not change in size, thus the volume and area integral are no longer integrating over a volume and area that are functions of time.

With the inclusion of the stress tensor for a Newtonian fluid (Equation 3.26), we get an **integral form for the compressible Navier–Stokes equations**:

$$\iiint_{\mathcal{V}} \frac{\partial\left(\rho\vec{V}\right)}{\partial t}d\mathcal{V} + \oiint_{A}\rho\vec{V}\otimes\vec{V}\cdot\vec{n}dA = \oiint_{A}\vec{\vec{T}}\cdot\vec{n}dA + \iiint_{\mathcal{V}}\rho\vec{g}d\mathcal{V}$$

$$\text{where} \quad \vec{\vec{T}} = -p\vec{\vec{I}} - \frac{2}{3}\mu\left(\vec{\nabla}\cdot\vec{V}\right)\vec{\vec{I}} + \mu\left(\vec{\nabla}\vec{V} + \left(\vec{\nabla}\vec{V}\right)^{\dagger}\right) \tag{4.13}$$

We can now take a couple more steps towards developing another differential equation version of the equations. Notice we have a situation where we have an area integral over a term that is dotted with an outward normal on both the left- and right-hand sides of the equal sign. We can use the divergence theorem on Equation 4.13 to get:

$$\iiint_{\mathcal{V}} \frac{\partial\left(\rho\vec{V}\right)}{\partial t}d\mathcal{V} + \iiint_{\mathcal{V}}\vec{\nabla}\cdot\left(\rho\vec{V}\otimes\vec{V}\right)d\mathcal{V} = \iiint_{\mathcal{V}}\vec{\nabla}\cdot\vec{\vec{T}}d\mathcal{V} + \iiint_{\mathcal{V}}\rho\vec{g}d\mathcal{V}$$

$$\text{where} \quad \vec{\vec{T}} = -p\vec{\vec{I}} - \frac{2}{3}\mu\left(\vec{\nabla}\cdot\vec{V}\right)\vec{\vec{I}} + \mu\left(\vec{\nabla}\vec{V} + \left(\vec{\nabla}\vec{V}\right)^{\dagger}\right). \tag{4.14}$$

Like before, since all terms are now integrals, we can "shrink down" to an infinitesimal volume (i.e., get rid of the volume integrals since we are integrating over an arbitrary volume) to give us:

$$\frac{\partial(\rho\vec{V})}{\partial t} + \vec{\nabla}\cdot\left(\rho\vec{V}\otimes\vec{V}\right) = \vec{\nabla}\cdot\vec{\vec{T}} + \rho\vec{g}$$

(4.15)

where $\vec{\vec{T}} = -p\vec{\vec{I}} - \frac{2}{3}\mu\left(\vec{\nabla}\cdot\vec{V}\right)\vec{\vec{I}} + \mu\left(\vec{\nabla}\vec{V} + \left(\vec{\nabla}\vec{V}\right)^{\dagger}\right)$.

Plugging the stress tensor into the equation leads to:

$$\frac{\partial(\rho\vec{V})}{\partial t} + \vec{\nabla}\cdot\left(\rho\vec{V}\otimes\vec{V}\right) = \vec{\nabla}\cdot\left(-p\vec{\vec{I}} - \frac{2}{3}\mu\left(\vec{\nabla}\cdot\vec{V}\right)\vec{\vec{I}} + \mu\left(\vec{\nabla}\vec{V} + \left(\vec{\nabla}\vec{V}\right)^{\dagger}\right)\right) + \rho\vec{g}.$$ (4.16)

Like we did in the last chapter, we can rewrite the right-hand side of Equation 4.16 using the gradients of pressure and the divergence of velocity to give us a form of the Navier–Stokes equations often known as the **compressible Navier–Stokes equations in conservation form**:

$$\boxed{\frac{\partial(\rho\vec{V})}{\partial t} + \vec{\nabla}\cdot\left(\rho\vec{V}\otimes\vec{V}\right) = -\nabla p - \vec{\nabla}\left(\frac{2}{3}\mu\vec{\nabla}\cdot\vec{V}\right) + \vec{\nabla}\cdot\left(\mu\vec{\nabla}\vec{V} + \mu\left(\vec{\nabla}\vec{V}\right)^{\dagger}\right) + \rho\vec{g}}$$

(4.17)

We can also use the Newtonian stress tensor for an incompressible flow, Equation 3.44, in Equation 4.12 to get the **integral form of the incompressible Navier–Stokes equations**:

$$\boxed{\begin{array}{l} \iiint_{\mathcal{V}}\frac{\partial(\rho\vec{V})}{\partial t}d\mathcal{V} + \oiint_{A}\rho\vec{V}\otimes\vec{V}\cdot\vec{n}dA = \oiint_{A}\vec{\vec{T}}\cdot\vec{n}dA + \iiint_{\mathcal{V}}\rho\vec{g}d\mathcal{V} \\[12pt] \text{where}\quad \vec{\vec{T}} = \vec{\vec{T}}_{incompressible} = -p\vec{\vec{I}} + \mu\left(\vec{\nabla}\vec{V} + \left(\vec{\nabla}\vec{V}\right)^{\dagger}\right) \end{array}}$$

(4.18)

In addition, we can use the divergence of the incompressible Newtonian stress tensor, Equation 3.50, in Equation 4.15 to get the **conservation form of the incompressible Navier–Stokes equations**:

$$\boxed{\frac{\partial(\rho\vec{V})}{\partial t} + \vec{\nabla}\cdot\left(\rho\vec{V}\otimes\vec{V}\right) = -\nabla p + \mu\nabla^{2}\vec{V} + \rho\vec{g}}$$

(4.19)

Whew, we have covered a lot of ground. In the next section, we are going to recap the main ideas thus far.

4.2 Take a Breath: Let's Review So Far

We have discussed the conservation of mass (leading to the continuity equation) and the conservation of momentum (leading to the Navier–Stokes equations). We discussed the idea of the two different descriptions of fluid flow in Chapter 2, one being a Lagrangian description (which follows a fluid element) and the other an Eulerian description (watching flow pass a region or point in space). These descriptions are related through the material derivative, i.e.:

$$\frac{D}{Dt} = \frac{\partial}{\partial t} + \vec{V} \cdot \vec{\nabla}$$

where the $\frac{D}{Dt}$ indicates the change of some property (such as temperature, pressure, density, velocity) of a moving fluid element (Lagrangian description). The right side (i.e., $\frac{\partial}{\partial t} + \vec{V} \cdot \vec{\nabla}$) considers the flow moving past a point in space (Eulerian description).

We have introduced four different forms, they are: 1. Lagrangian form, 2. Non-conservation form, 3. Conservation form, and 4. Integral form. In addition, for each form of the Navier–Stokes equations, we used a stress tensor for a Newtonian fluid for either a compressible or an incompressible flow.

The Lagrangian, non-conservation, and conservation forms are all written as complex partial differential equations while the integral form is an integro-differential equation since it involves both integrals and derivatives. In Chapter 5, we will cover an additional equation, the energy equation, which applies the first law of thermodynamics to a moving fluid. The energy equation can also be written in these four forms.

Let's touch on each of these forms briefly.

4.2.1 Lagrangian Form

The continuity and the Navier–Stokes equations can be summarized in Lagrangian form in Table 4.1.

This form of the equations follows a fluid element. While this form is the most compact and, in some ways, the most understandable from a typical mechanics point of view, its usage in analysis can be somewhat limiting. However, there are a number of powerful computational techniques that use this form as the basis for their computer programs. In general, computer programs that solve the fluid mechanics equations make up what is called computational fluid dynamics, or CFD for short. Considering the complexity of the equations, by hand solutions (i.e., solving problems using the pen and paper techniques of calculus and differential equations) are few and far between. Thus, the use of

Table 4.1 *Lagrangian form of the continuity and Navier–Stokes equations (Energy equation given in Equation 5.42)*

Compressible version

Continuity Equation:

$$\frac{D\rho}{Dt} + \rho\vec{\nabla} \cdot \vec{V} = 0$$

Navier–Stokes:

$$\rho\frac{D\vec{V}}{Dt} = -\vec{\nabla}p - \vec{\nabla}\left(\tfrac{2}{3}\mu\left(\vec{\nabla} \cdot \vec{V}\right)\right) + \vec{\nabla} \cdot \left(\mu\left(\vec{\nabla}\vec{V} + \left(\vec{\nabla}\vec{V}\right)^{\dagger}\right)\right) + \rho\vec{g}$$

Incompressible version

Continuity (incompressibility constraint):

$$\vec{\nabla} \cdot \vec{V} = 0$$

Navier–Stokes:

$$\rho\frac{D\vec{V}}{Dt} = -\vec{\nabla}p + \mu\nabla^2\vec{V} + \rho\vec{g}$$

---

computers to solve fluids problems is what is quite often done. One CFD method that uses the Lagrangian form as a starting point is called smoothed-particle hydrodyamics (SPH). These programs model a moving fluid as a collection of particles (i.e., fluid elements) and calculates the force on the fluid elements based off the right-hand side of the equations in Table 4.1. Computational fluid dynamics methods, including the SPH technique, are not within the scope of this book. Nevertheless, they make up a significant portion of fluid mechanics research and engineering work in modern times. In general, however, the Lagrangian form is probably not used as often as other forms of the equations.

4.2.2 Non-Conservation Form

If we were to expand out the material derivative from the Lagrangian description to the Eulerian description we get what is called the non-conservation form. The non-conservation form of the continuity equation and the Navier–Stokes equations are given in Table 4.2.

These are the most common forms of the Navier–Stokes equations, especially the incompressible version. In fact, many times when one mentions the Navier–Stokes equations, they are referencing the non-conservation form of the incompressible Navier–Stokes equations. This form is studied extensively

Table 4.2 *Non-conservative form of the continuity and Navier–Stokes equations (Energy equation given in Equation 5.41)*

Compressible version

Continuity Equation:

$$\frac{\partial \rho}{\partial t} + \vec{V} \cdot \vec{\nabla} \vec{V} + \rho \vec{\nabla} \cdot \vec{V} = 0$$

Navier–Stokes:

$$\rho \left(\frac{\partial \vec{V}}{\partial t} + \vec{V} \cdot \vec{\nabla} \vec{V} \right) = -\vec{\nabla} p - \vec{\nabla} \left(\tfrac{2}{3} \mu \left(\vec{\nabla} \cdot \vec{V} \right) \right) + \vec{\nabla} \cdot \left(\mu \left(\vec{\nabla} \vec{V} + \left(\vec{\nabla} \vec{V} \right)^{\dagger} \right) \right) + \rho \vec{g}$$

Incompressible version

Continuity (incompressibility constraint):

$$\vec{\nabla} \cdot \vec{V} = 0$$

Navier–Stokes:

$$\rho \left(\frac{\partial \vec{V}}{\partial t} + \vec{V} \cdot \vec{\nabla} \vec{V} \right) = -\vec{\nabla} p + \mu \nabla^2 \vec{V} + \rho \vec{g}$$

---

in fluid mechanics classes, not only in physics and engineering, but also in certain areas of mathematics. Interestingly enough, however, the compressible continuity equation is usually not written in this form and is instead often written in conservation form, even when seen alongside the non-conservation form of the Navier–Stokes equations.

In this form, unlike Lagrangian form, the velocity (as well as pressure and density) are property values at given points in space. That is, they are fields. In Lagrangian form, the property values follow the fluid elements.

We will spend a good deal of time on the incompressible version of the equations in non-conservation form.

4.2.3 Conservation Form

Another form of the equations is the conservation form, which is sometimes called the Eulerian form. This form was developed by assuming an Eulerian framework as a starting point and tracking fluxes of either mass or momentum past a fixed control volume in space. The conservation form is given in Table 4.3.

Table 4.3 *Conservative form of the continuity and Navier–Stokes equations (Energy equation given in Equation 5.39)*

Compressible version

Continuity Equation:

$$\frac{\partial \rho}{\partial t} + \vec{\nabla} \cdot \left(\rho \vec{V} \right) = 0$$

Navier–Stokes:

$$\frac{\partial \left(\rho \vec{V} \right)}{\partial t} + \vec{\nabla} \cdot \left(\rho \vec{V} \otimes \vec{V} \right) = -\nabla p - \vec{\nabla} \left(\tfrac{2}{3} \mu \vec{\nabla} \cdot \vec{V} \right) + \vec{\nabla} \cdot \left(\mu \vec{\nabla} \vec{V} + \mu \left(\vec{\nabla} \vec{V} \right)^{\dagger} \right) + \rho \vec{g}$$

Incompressible version

Continuity (incompressibilty constraint):

$$\vec{\nabla} \cdot \vec{V} = 0$$

Navier–Stokes:

$$\frac{\partial \left(\rho \vec{V} \right)}{\partial t} + \vec{\nabla} \cdot \left(\rho \vec{V} \otimes \vec{V} \right) = -\vec{\nabla} p + \mu \nabla^2 \vec{V} + \rho \vec{g}$$

This form is also fairly common, especially in engineering. Engineering analysis, particularly engineering thermodynamic analysis, relies heavily on fixed control volumes. Think of a fluid passing through a nozzle of a turbine engine on a jet airplane. Engineers sometimes like to treat these devices (nozzle, turbine, etc.) as fixed volumes in space and analyze the flow in and out of them. Thus, thinking of a fluid passing through a volume of space makes this form somewhat more natural for an engineer. In addition, many of the CFD programs that are developed, at least for engineering uses, consider the fluid flow from an Eulerian perspective. One of the exceptions to this is the smoothed-particle hydrodynamics programs discussed earlier, which uses a Lagrangian form.

4.2.4 Integral Form

The integral form of the equations is given in Table 4.4. Like the conservation form, the integral form is also somewhat common since it was developed from an Eulerian (or control volume) perspective, which is familiar to engineers. In addition, the integral form is also a very common starting point for the

Table 4.4 *Integral form of the continuity and Navier–Stokes equations (Energy equation given in Equation 5.38)*

Continuity Equation:

$$\iiint_{\mathcal{V}} \frac{\partial \rho}{\partial t} d\mathcal{V} + \oiint_{A} \rho \vec{V} \cdot \vec{n} dA = 0$$

Navier–Stokes:

$$\iiint_{\mathcal{V}} \frac{\partial (\rho \vec{V})}{\partial t} d\mathcal{V} + \oiint_{A} \rho \vec{V} \otimes \vec{V} \cdot \vec{n} dA = \oiint_{A} \vec{\vec{T}} \cdot \vec{n} dA + \iiint_{\mathcal{V}} \rho \vec{g} d\mathcal{V}$$

where

$$\vec{\vec{T}} = -p\vec{\vec{I}} - \tfrac{2}{3}\mu \left(\vec{\nabla} \cdot \vec{V}\right)\vec{\vec{I}} + \mu \left(\vec{\nabla}\vec{V} + \left(\vec{\nabla}\vec{V}\right)^{\dagger}\right) \quad \text{(compressible)}$$

or

$$\vec{\vec{T}} = -p\vec{\vec{I}} + \mu \left(\vec{\nabla}\vec{V} + \left(\vec{\nabla}\vec{V}\right)^{\dagger}\right) \quad \text{(incompressible)}$$

---

algorithms used in a computational fluid dynamics program. In fact, it may be the most common starting point for CFD programs. The reason for this has to do with the fact that it is sometimes easier to deal with complex geometries with the integral form as a starting point.

4.3 Incompressible Equations in 2D Cartesian Coordinates

One of the more useful and easiest versions of the governing equations for tackling problems is the incompressible equations in two-dimensional Cartesian coordinates. In particular, the non-conservation form of the incompressible Navier–Stokes equations.

Writing out the non-conservation form of the incompressible Navier–Stokes equations, using the del notation, we get:

$$\vec{\nabla} \cdot \vec{V} = 0$$

$$\rho \left(\frac{\partial \vec{V}}{\partial t} + \vec{V} \cdot \vec{\nabla}\vec{V} \right) = -\vec{\nabla}p + \mu \nabla^2 \vec{V} + \rho \vec{g}$$

Expanding out into Cartesian coordinates (from Equations 1.43, 2.13, 3.51) leads to (with gravity in the y-direction):

Continuity:

$$\frac{\partial u}{\partial x} + \frac{\partial v}{\partial y} + \frac{\partial w}{\partial z} = 0 \tag{4.20}$$

The incompressible Navier–Stokes broken out in x-, y-, and z-directions:

$$\rho \left(\frac{\partial u}{\partial t} + u\frac{\partial u}{\partial x} + v\frac{\partial u}{\partial y} + w\frac{\partial u}{\partial z} \right) = -\frac{\partial p}{\partial x} + \mu \left(\frac{\partial^2 u}{\partial x^2} + \frac{\partial^2 u}{\partial y^2} + \frac{\partial^2 u}{\partial z^2} \right) \tag{4.21}$$

$$\rho \left(\frac{\partial v}{\partial t} + u\frac{\partial v}{\partial x} + v\frac{\partial v}{\partial y} + w\frac{\partial v}{\partial z} \right) = -\frac{\partial p}{\partial y} + \mu \left(\frac{\partial^2 v}{\partial x^2} + \frac{\partial^2 v}{\partial y^2} + \frac{\partial^2 v}{\partial z^2} \right) + \rho g_y \tag{4.22}$$

$$\rho \left(\frac{\partial w}{\partial t} + u\frac{\partial w}{\partial x} + v\frac{\partial w}{\partial y} + w\frac{\partial w}{\partial z} \right) = -\frac{\partial p}{\partial z} + \mu \left(\frac{\partial^2 w}{\partial x^2} + \frac{\partial^2 w}{\partial y^2} + \frac{\partial^2 w}{\partial z^2} \right). \tag{4.23}$$

For a two dimensional scenario, we will ignore the z-dimension. Thus, we can write the equations as (with some labeling of terms):

$$\frac{\partial u}{\partial x} + \frac{\partial v}{\partial y} = 0 \tag{4.24}$$

$$\rho \left(\underbrace{\frac{\partial u}{\partial t}}_{\substack{\text{local time} \\ \text{derivative}}} + \underbrace{u\frac{\partial u}{\partial x} + v\frac{\partial u}{\partial y}}_{\substack{\text{advective} \\ \text{transport} \\ \text{(inertia term)}}} \right) = \underbrace{-\frac{\partial p}{\partial x}}_{\substack{\text{pressure} \\ \text{gradient}}} + \underbrace{\mu \left(\frac{\partial^2 u}{\partial x^2} + \frac{\partial^2 u}{\partial y^2} \right)}_{\substack{\text{diffusive transport} \\ \text{(viscous force)}}} \tag{4.25}$$

$$\rho \left(\underbrace{\frac{\partial v}{\partial t}}_{\substack{\text{local time} \\ \text{derivative}}} + \underbrace{u\frac{\partial v}{\partial x} + v\frac{\partial v}{\partial y}}_{\substack{\text{advective} \\ \text{transport} \\ \text{(inertia term)}}} \right) = \underbrace{-\frac{\partial p}{\partial y}}_{\substack{\text{pressure} \\ \text{gradient}}} + \underbrace{\mu \left(\frac{\partial^2 v}{\partial x^2} + \frac{\partial^2 v}{\partial y^2} \right)}_{\substack{\text{diffusive transport} \\ \text{(viscous force)}}} + \underbrace{\rho g_y}_{\substack{\text{gravity} \\ \text{body force}}} \tag{4.26}$$

The two-dimensional version of the equations will turn out to be very useful for problems and will be the "goto" version of the equations that we will use. Equation 4.24 is the 2D continuity equation for incompressible flows. Equations 4.25 and 4.26 make up the 2D incompressible Navier–Stokes equations, with Equation 4.25 being the x-momentum equation and Equation 4.26 is the y-momentum equation.

4.4 Boundary Conditions

As you may recall from your differential equations course, when a differential equation is solved it typically involves a general solution. A **general solution** is one that satisfies the differential equation but requires more information (in the form of initial and boundary conditions) in order to be a solution for a particular problem. General solutions usually contain constants of integration. These constants arise from the integration procedures when solving a given differential equation. The constants are determined by applying either boundary or initial conditions for a given problem. **Boundary conditions** are typically values of the dependent variable or values of the derivatives of the dependent variable at a given spatial location. An example boundary condition could be written as:

$$\text{at } x = 1 \text{ m}, \quad u = 10 \text{ m/s}.$$

Another way of writing the same boundary condition (which is sometimes more convenient) is:

$$u\Big|_{x=1 \text{ m}} = 10 \text{ m/s}.$$

This boundary condition is stating that the value of the x-velocity (i.e., u) is 10 m/s at the x-location of 1 meter.

An **initial condition** is a value given to a dependent variables (or its derivatives) at a given time. As an example:

$$\text{at } t = 0 \text{ s}, \quad v = 0 \text{ m/s}.$$

In general, the number of boundary conditions (or initial conditions) needed for a differential equation often matches the highest order derivative of a given independent variable, assuming the geometry is simple (such as a square or a cube in Cartesian coordinates). For example, consider the two-dimensional incompressible momentum equation in the x-direction, that is, Equation 4.25. In Equation 4.25, the highest order derivative of u with respect to x is the second order derivative in the diffusive term $\left(\text{i.e., } \frac{\partial^2 u}{\partial x^2}\right)$, thus we generally need to consider two boundary conditions in the x-direction. Since the highest order derivative of u with respect to y is also second order, we would need two boundary conditions in the y-direction. The time derivative in Equation 4.25 is only first order $\left(\frac{\partial u}{\partial t}\right)$, thus a single initial condition based off of time is usually all that is necessary.

In general, there are three main types of boundary conditions:

(i) **Dirichlet boundary condition**: This type of boundary condition occurs when a specific value for the dependent variable (i.e., the velocity in the case of the Navier–Stokes equations) is set at a given boundary. An example is the following:

$$\text{at } y = 0.1 \text{ m}, \quad v = 1 \text{ m/s}.$$

The boundary condition above sets the value of the y-velocity to 1 m/s whenever $y = 0.1$ m. The Dirichlet boundary condition is very common in fluid mechanics. In fact, the most useful and important boundary condition in fluid mechanics is called the **no-slip boundary condition**, which is a Dirichlet type of boundary condition. The no-slip condition occurs when the velocity of the fluid is set to be the same as the velocity of the surface (usually a solid surface) on which the fluid is in contact. We will utilize the no-slip condition shortly in an example. Physically, the no-slip condition stems from molecules of a solid surface slowing down the molecules of the fluid passing over it. If we consider a large enough number of molecular collisions with the solid surface and the fluid, the fluid is considered to be essentially "stuck" at the solid surface due to these molecular collisions.

(ii) **Neumann boundary condition**: This type of boundary condition specifies the value of a derivative of the dependent variable at a specified boundary point. An example is:

$$\mu \frac{\partial u}{\partial y}\Big|_{y=0} = 10 \text{ Pa}.$$

This boundary condition is essentially setting the shear stress at $y = 0$ to be 10 Pa. Neumann boundary conditions are used in fluid mechanics when we have a specified stress value on a surface. The reason for this is because, as you may recall, the stress on a surface is related to the derivatives of the velocity. For example, the stress in the y-direction on the x-face of a Cartesian element is $\tau_{xy} = \mu\left(\frac{\partial u}{\partial y} + \frac{\partial v}{\partial x}\right)$. If $v = 0$, we would be left with just $\tau_{xy} = \mu\frac{\partial u}{\partial y}$. Note, a more general version of this type of boundary condition would be (as an example): at $y = 0$, $\vec{\tau} \cdot \vec{n} = 10$ Pa, where $\vec{n}$ is the unit normal of the boundary of interest.

(iii) **Robin boundary condition**: This type of boundary condition utilizes both the value of the dependent variable in question as well as its derivatives. An example would be:

$$\mu \frac{\partial u}{\partial y}\Big|_{y=0} = 10u\Big|_{y=0}$$

In the next section, we will take a look at two simple, yet very common, examples of the Navier–Stokes equations in practice and the use of applying boundary conditions to the general solution to obtain a particular solution for a given problem.

4.5 Examples

4.5.1 Couette, or Shear-Driven, Flow

A very standard first example for solving a simplified version of the Navier–Stokes equations is a shear-driven flow scenario known as Couette flow. In Couette flow, fluid sits between two parallel plates as shown in 4.1. One of the plates is moved to the left or right, causing the fluid to move. In the example given in Figure 4.1, the top plate is being pulled to the right with a velocity, U_T. If we assume that there is no pressure gradient driving the fluid (i.e., $\vec{\nabla} p = 0$), then the only way the fluid can move between the two plates is by a friction force via the diffusive transport term. The way to think of this friction force is the same way we thought of the no-slip condition, that is to envision the atoms of the fluid adjacent to the plate bumping into the atoms of the plate. If the plate moves to the right at U_T, then the average velocity of the atoms of the plate (assuming a large number of atoms to be able to assume a continuum) in the x-direction will move to the right at U_T. The atoms of the fluid adjacent to the plate will (on average) be pushed in the same direction as the plate with the

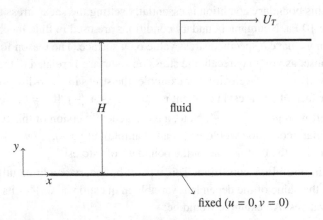

Figure 4.1 Schematic of the Couette flow setup. Fluid is contained between two parallel plates (shown here in 2D) and is driven to flow by moving one of the plates to the right or left (in this scenario, the top plate is moving to the right with a velocity of U_T).

same velocity. Thus the flow velocity at the top plate will be U_T. This is the essence of the no-slip condition.

Let's **solve for the steady-state (i.e., no time variation) velocity profile of the fluid between the plates** (profile usually means finding the velocity as a function of the up and down direction, i.e., as a function of y). To do so, the best strategy is to take each governing equation (e.g., mass, Navier–Stokes) and make some simplifications.

Our first simplification[1] will be to assume our starting point equations will be the two-dimensional version of the incompressible flow equations. Our two-dimensional incompressible continuity equation is (from Equation 4.24):

$$\frac{\partial u}{\partial x} + \frac{\partial v}{\partial y} = 0.$$

The next assumption we are going to make is that the velocity of the flow will only be in the x-direction. Therefore v will be zero. The resulting simplified continuity equation is now just the derivative of the x-velocity (u) with respect to x is 0 (since v goes away):

$$\frac{\partial u}{\partial x} = 0. \tag{4.27}$$

Equation 4.27 implies that u does not change with x. However, it can still change with y since only the derivative with respect to x is zero. A consequence of the derivative of u with respect to x being zero is that the u profile (i.e., u as a function of y) will be the same at various cross-sections of the channel, as shown in Figure 4.2. Such a flow (one where the velocity does not change downstream) is called a **fully-developed flow**.

The next step in solving for the velocity profile is to move onto the Navier–Stokes equations and simplify. We can start with the incompressible Navier–Stokes equations in two dimensions, i.e., Equations 4.25 and 4.26. We can make some additional assumptions about our problem so that we can eliminate more terms. The first assumption we are going to make is that the flow is steady (i.e., all time derivatives are zero). We are also going to assume that there are no body forces ($\rho g_y = 0$) and that the driving force for the flow is not pressure but shear $\left(\text{therefore the pressure derivatives are zero, } \frac{\partial p}{\partial x} = 0 \text{ and } \frac{\partial p}{\partial y} = 0\right)$. In addition, our continuity result previously indicated the flow is also fully developed $\left(\text{i.e., } \frac{\partial u}{\partial x} = 0\right)$. We can now cross off the terms in the x-momentum portion of the Navier–Stokes equations (Equation 4.25):

---

[1] Actually, our first assumption is to assume something called laminar flow, as opposed to turbulent flow. Turbulent flow is a completely different ballgame and even though the Navier–Stokes equations (the 3D version) govern turbulent flow as well, we are going to stick with simple laminar flow. A discussion on turbulence will come in Chapter 6.

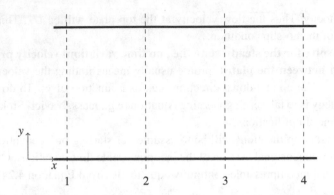

Figure 4.2 The velocity of the flow passing through each cross-section (labeled 1, 2, 3, 4) will be the same in a fully-developed flow and would continue to be the same as we progress further downstream.

$$\rho\left(\frac{\partial u}{\partial t} + u\frac{\partial u}{\partial x} + v\frac{\partial u}{\partial y}\right) = -\frac{\partial p}{\partial x} + \mu\left(\frac{\partial^2 u}{\partial x^2} + \frac{\partial^2 u}{\partial y^2}\right). \tag{4.28}$$

The y-momentum portion of the Navier–Stokes equations, Equation 4.26, is rather boring since we essentially cross off all terms. Thus, we are really only concerning ourselves with the x-direction momentum.

Therefore, we are left with the following very simple momentum equation (after dividing out the μ):

$$0 = \frac{d^2 u}{dy^2} \tag{4.29}$$

This is the governing equation for Couette flow. We switched to an ordinary derivative since u is only a function of y. The solution to Equation 4.29 is a simple line equation:

$$u = C_1 y + C_2 \tag{4.30}$$

where C_1 and C_2 are constants of integration. Equation 4.30 is the general solution for this problem. In order to find values for C_1 and C_2, boundary conditions need to be applied.

The no-slip boundary condition for u is utilized at the top and bottom plates. Recall that the no-slip boundary condition implies that the velocity of the fluid is the same as the velocity of the adjacent solid surface. Thus, in mathematical form, the boundary conditions are:

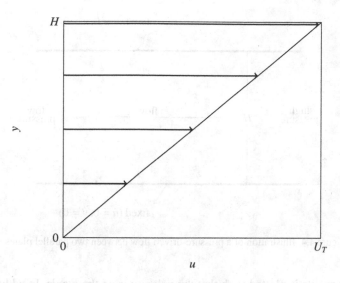

Figure 4.3 Velocity profile for Couette flow (assuming the top plate is moving to the right).

$$\text{at } y = 0, \quad u = 0$$

$$\text{at } y = H, \quad u = U_T.$$

Applying the first boundary condition (i.e., at $y = 0$, $u = 0$) to our general solution, Equation 4.30, leads to:

$$0 = C_1 0 + C_2$$

$$\therefore C_2 = 0.$$

Note the $\therefore$ means therefore. Next apply the second boundary condition (i.e., at $y = H$, $u = U_T$) to our general solution, i.e.:

$$U_T = C_1 H + \cancel{C_2}^{\,0 \text{ from first boundary condition}}$$

$$\therefore C_1 = \frac{U_T}{H}$$

The solution for the x-velocity profile is now obtained by plugging in the expressions for C_1 and C_2 into Equation 4.30:

$$u = \frac{U_T}{H} y \tag{4.31}$$

which is just a straight line starting at zero velocity and linearly going to the top plate velocity at U_T. The profile can be seen in Figure 4.3. Notice that the

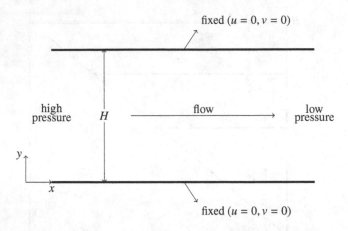

Figure 4.4 Illustration of a pressure-driven flow between two parallel plates.

velocity profile is plotted such that the velocity is on the x-axis. In addition, it is common to draw arrows denoting velocity vectors in a velocity profile plot (as shown in Figure 4.3).

4.5.2 Pressure-driven Flow

The next problem we are going to study is pressure-driven flow between two parallel plates. Typically, pressure-driven flow is studied in pipes using cylindrical coordinates. Such a flow is called Poiseuille flow. However, we will just stick to pressure-driven flow between parallel plates in order to utilize Cartesian coordinates. Like Couette flow, we are going to calculate a velocity profile between the plates.

To get started, consider the scenario given in Figure 4.4. The flow in this scenario is driven by a constant pressure gradient in the x-direction. The assumptions we are going to make are as follows: incompressible flow ($\vec{\nabla} \cdot \vec{V} = 0$), steady state (all time derivatives are zero), the flow is only in the x-direction ($v = w = 0$), any gradients in the z-direction are completely ignored (in other words, we will consider only the two-dimensional version of the equations), and gravitational effects are negligible ($\rho \vec{g} = 0$).

Considering all of these assumptions, we can make the following simplification to the two dimensional continuity equation:

$$\frac{\partial u}{\partial x} + \underbrace{\frac{\partial v}{\partial y}}_{v=0} = 0$$

which leads to our fully developed situation, i.e.:

$$\frac{\partial u}{\partial x} = 0$$

Since the gradients in our z-direction are assumed to be negligible, we are only left with the conclusion that either $u = 0$ everywhere or $u = f(y)$. Since $u = 0$ is clearly not the situation, then u must be a function of y.

The x-momentum of the two-dimensional incompressible Navier–Stokes equations becomes:

$$\rho\left(\frac{\partial u}{\partial t} + u\frac{\partial u}{\partial x} + v\frac{\partial u}{\partial y}\right) = -\frac{\partial p}{\partial x} + \mu\left(\frac{\partial^2 u}{\partial x^2} + \frac{\partial^2 u}{\partial y^2}\right) + f_{bx}.$$

This gives us:

$$0 = -\frac{dp}{dx} + \mu\frac{d^2 u}{dy^2}$$

Note we are again replacing the partial derivatives with ordinary ones since u is only a function of y. Rearranging we get:

$$\frac{d^2 u}{dx^2} = \frac{1}{\mu}\frac{dp}{dx}. \tag{4.32}$$

Equation 4.32 is the governing equation for u for this problem. For the y-momentum, which is much less interesting, we have $\frac{\partial p}{\partial y} = 0$ since everything else cancels. This implies that pressure is not a function of y and only a function of x, as we could have guessed.

Solving for u yields a general solution of:

$$u = \frac{1}{2\mu}\frac{dp}{dx}y^2 + C_1 y + C_2 \tag{4.33}$$

We now need to apply the following boundary conditions:

$$\text{at } y = 0, \quad u = 0$$
$$\text{at } y = H, \quad u = 0.$$

Applying the boundary conditions will allow us to obtain solutions for C_1 and C_2:

$$\text{at } y = 0, \quad u = 0 \quad \rightarrow 0 = \frac{1}{2\mu}\frac{dp}{dx}0^2 + C_1 0 + C_2 \quad \Rightarrow C_2 = 0$$

$$\text{at } y = H, \quad u = 0 \quad \rightarrow 0 = \frac{1}{2\mu}\frac{dp}{dx}H^2 + C_1 H + C_2^{\,0} \quad \Rightarrow C_1 = -\frac{1}{2\mu}\frac{dp}{dx}H.$$

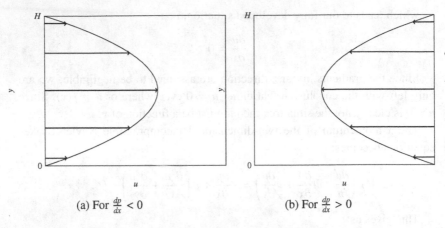

(a) For $\frac{dp}{dx} < 0$ (b) For $\frac{dp}{dx} > 0$

Figure 4.5 Velocity profile for pressure-driven flow.

Plugging C_1 and C_2 into the general solution for the velocity profile leads to:

$$u = \frac{1}{2\mu}\frac{dp}{dx}\left(y^2 - Hy\right) \tag{4.34}$$

Equation 4.34 is plotted on a y vs. u curve in Figure 4.5. Like the Couette flow example velocity profile in Figure 4.3, arrows are drawn to illustrate velocity vectors.

Problems

4.1   Find $\vec{A} \otimes \vec{B}$ if $\vec{A} = 3\hat{i} + 5\hat{j} - 10\hat{k}$ and $\vec{B} = 5\hat{i} - 2\hat{j} - \hat{k}$.

4.2   Determine $\oiint_A \rho\vec{V} \otimes \vec{V} \cdot \vec{n}dA$ for a 1 cm × 1 cm × 1 cm Cartesian element if $\vec{V} = 5\sqrt{x}\,\hat{i}$ m/s and the density is 1000 kg/m^3.

4.3   Consider incompressible flow between two parallel plates a distance, H, apart. If the bottom plate moves with a velocity of U_B and the top is fixed, find an expression of the velocity profile assuming there is no pressure gradient.

4.4   Find an expression for the shear stress on the bottom and top plates from Problem 4.3.

4.5   Consider an incompressible flow between two parallel plates a distance, $H = 0.5$ cm, apart. The top plate is moving with a velocity to the right of 1 m/s. There is also a constant pressure gradient that is resisting the flow in the x-direction with a magnitude of 2 pascals per meter. The dynamic viscosity of the fluid is 2×10^{-5} Pa·s. Determine the velocity at the midpoint of the parallel plates.

4.6 In your own words, describe the various forms of the Navier–Stokes equations discussed in this book and how they were obtained.

4.7 Show that the conservation form and the non-conservation form of the Navier–Stokes equations are equivalent. You may use Cartesian coordinates

5

The Energy Equation and a Discussion on Diffusion and Advection

As we have seen, the Navier–Stokes equations require a separate mass continuity equation. There is yet another very important equation that is often employed alongside the Navier–Stokes equations: the energy equation. The energy equation is required to fully describe compressible flows. However, it is typically not required in order to account for the dynamics of an incompressible flow. Nevertheless, even for incompressible flows, it is still often useful to study alongside the Navier–Stokes equations. For instance, the energy equation will share two terms in common with the Navier–Stokes equations: the advective transport term and the diffusive transport term. Studying these two terms in the context of the energy equation will provide some guidance in interpreting the Navier–Stokes equations.

It should be stated up front that the energy equation can look intimidating. We will derive the energy equation but a complete understanding of it involves a good background in thermodynamics. While it is assumed you have had at least some exposure to thermodynamics, an attempt will be made to cover the essentials needed to get a grasp of the basic concepts. Even so, we will not go through all of the details of the thermodynamics. In addition, most of our discussion will be on a much simpler version of the energy equation. Namely, the energy equation for incompressible flows.

5.1 Conservation of Energy

The energy equation is derived starting from the first law of thermodynamics, which is a statement of the conservation of energy. The standard equation associated with the first law of thermodynamics relates the change in energy of a system (ΔE) with the energy transfer leaving or entering the system via heat (Q) and work (W) in the following manner:

$$\Delta E = Q - W. \tag{5.1}$$

The ΔE can be written as the final energy of the system, E_f, minus the initial energy of the system, E_i. Therefore, Equation 5.1 becomes:

$$E_f - E_i = Q - W. \tag{5.2}$$

Energy is exchanged between the system and its surroundings. The **surroundings** of a system is everything outside the system. The mechanisms by which energy is transferred are heat and work. Given that heat and work transfer energy to and from a system, we need to have a convention for when energy is transferred to a system and when it is transferred out of a system. The convention for the energy transfer by heat is the following:

- $Q > 0$: Energy is added to the system by heat (i.e., energy is taken away from the surroundings).
- $Q < 0$: Energy is taken away from the system by heat (i.e., energy is given to the surroundings).

The sign convention used for the work term is the following:[1]

- $W < 0$: Energy is added to the system by work. This is sometimes written as the work done on the system.
- $W > 0$: Energy is taken away from the system by work. This is often written as the work done by the system.

Equation 5.1 is the most commonly cited version of the first law of thermodynamics. Each term has units of energy, that is, joule (J) or newton-meter (N·m). It will come in handy when discussing the various components of energy, as we will do later. However, a slightly different form of energy conservation is typically used when applied to a moving fluid. In that case, one typically sets up the energy balance much as we discussed in Chapter 1, namely:

the change of energy in a system in a given unit of time	=	time rate of energy entering the system	−	time rate of energy leaving the system	+	time rate of a source(sink) of energy inside the system

The rate of energy leaving or entering the system can come from heat, it can come from work, or it can come from fluid flow passing through a control

---

[1] Some books, notably in chemistry, use a flipped sign convention for work, in which case the first law would be $\Delta E = Q + W$.

volume (in the case of an Eulerian description). In symbolic form, we can write the conservation of energy principle as:

$$\frac{dE_{sys}}{dt} = \dot{E}_{in} - \dot{E}_{out} + \dot{Q} - \dot{W} + Source \qquad (5.3)$$

where E_{sys} is the energy in our system, $\dot{Q}$ is the rate of energy transfer by heat either to or from the system, $\dot{W}$ is the rate of work (i.e., power) being done by or on the system, and *Source* is the rate of energy generation within the system.[2] In addition, the $\dot{E}_{in}$ is the rate of energy from fluid flow coming into the system (i.e., control volume) and $\dot{E}_{out}$ is the rate of energy from fluid flow going out of the system/control volume. If we are using a Lagrangian perspective, the $\dot{E}$ terms will go away and we would be left with:

$$\frac{dE_{sys}}{dt} = \dot{Q} - \dot{W} + Source.$$

We will use an Eulerian approach for developing the energy equation, so we will keep the energy flow rates (i.e., the $\dot{E}$ terms).

The units of Equation 5.3 are no longer that of energy but a rate of energy, that is, joules per second or watts. The sign convention for rate of heat and power is the same as the sign convention for heat and work, that is:

- $\dot{Q} > 0$: Rate of energy is added to the system by heat
- $\dot{Q} < 0$: Rate of energy is taken away from the system by heat

- $\dot{W} < 0$: Rate of energy added to the system by work (i.e., power done on the system).
- $\dot{W} > 0$: Rate of energy taken away from the system by work (i.e., power done by the system).

Much like before when we did momentum conservation, we can define the energy within the control volume to be generally defined as density multiplied by an energy per mass integrated over the volume region of our fixed control volume, that is:

$$E_{sys} = \iiint_{\mathcal{V}} \rho e d\mathcal{V} \qquad (5.4)$$

where e is an energy per mass term known as the specific energy and has units of joules per kilogram (J/kg). Specific energy is a thermodynamic property of the system. A **property** is a quantity used to describe some aspect of a system. Thermodynamic properties have a thermodynamic origin (i.e., related to the laws of thermodynamics in some way) and include pressure, density, and

---

[2] Note a source term is not traditionally included in the "standard" equation used for the first law given by Equation 5.1, although it can be.

temperature, to name a few. Thermodynamic properties come in two flavors: extensive and intensive. An **extensive property** is a property that is dependent on the size of the system, such as energy and mass. An **intensive property**[3] is one that is independent of size, such as temperature, pressure, and density. Since specific energy is energy per mass, it is independent of size and is considered an intensive property. In general, whenever the word "specific" precedes the name of a property it implies that the property is taken on a per mass basis and is intensive.

Plugging Equation 5.4 into Equation 5.3 gives us:

$$\frac{d}{dt}\left(\iiint_V \rho e d\mathcal{V}\right) = \dot{E}_{in} - \dot{E}_{out} + \dot{Q} - \dot{W} + Source$$

$$\rightarrow \iiint_V \frac{\partial(\rho e)}{\partial t} d\mathcal{V} = \dot{E}_{in} - \dot{E}_{out} + \dot{Q} - \dot{W} + Source. \tag{5.5}$$

We are able to easily bring the time derivative into integral due to the fact that we are dealing with a control volume that is fixed in space.

An equation for the energy flow rate is based very much on the same concept we used for mass and momentum. We can write the energy flow rate in minus the energy flow rate out as an area integral of an energy flux ($\vec{E}''$) term dotted with the outward normal, that is:

$$\dot{E}_{in} - \dot{E}_{out} = -\oiint_A \vec{E}'' \cdot \vec{n} dA.$$

The units of energy flux, just like the units of mass and momentum flux, are going to be that of an energy per area per unit time, or $\frac{J}{m^2 s}$. Another way to think of energy flux is an energy per volume (written as ρe) multiplied by velocity. The equation for energy flux is the following:

$$\vec{E}'' = \rho e \vec{V}. \tag{5.6}$$

This means that the energy flow rate in minus out of the system is:

$$\dot{E}_{in} - \dot{E}_{out} = -\oiint_A \left(\rho e \vec{V}\right) \cdot \vec{n} dA. \tag{5.7}$$

Plugging Equation 5.7 into Equation 5.5 and bringing the energy flow rate terms to the left-hand side of the equal sign gives us a good starting point for the energy equation:

$$\boxed{\iiint_V \frac{\partial(\rho e)}{\partial t} d\mathcal{V} + \oiint_A \left(\rho e \vec{V}\right) \cdot \vec{n} dA = \dot{Q} - \dot{W} + Source} \tag{5.8}$$

---

[3] These are also sometimes called state variables in thermodynamics textbooks.

Labeling some of the terms might be useful:

$$\underbrace{\iiint_V \frac{\partial (\rho e)}{\partial t} V}_{\substack{\text{change of energy} \\ \text{with respect to time} \\ \text{in a fixed control volume}}} + \underbrace{\oiint_A (\rho e \vec{V}) \cdot \vec{n} dA}_{\text{energy flow rate}} = \underbrace{\dot{Q}}_{\text{rate of heat}} - \underbrace{\dot{W}}_{\text{power}} + \underbrace{Source}_{\substack{\text{rate of energy source} \\ \text{within control volume}}}$$

We will now begin looking at the individual pieces, starting with the general concept of energy in fluid mechanics.

5.1.1 Energy in Fluid Mechanics

Energy in fluid mechanics is usually broken up into three different types: internal energy ($\hat{U}$), kinetic energy (KE), and potential energy (PE). Simply put, $E = \hat{U} + KE + PE$. However, as we shall see, we will include the potential energy (PE) as a work term instead of an energy term. Thus, for us, the energy is just a sum of the internal energy and the kinetic energy of the system, that is, $E = \hat{U} + KE$. The kinetic energy is the "energy of motion" given by $KE = \frac{1}{2} m \vec{V} \cdot \vec{V}$, where m is the mass of the system. Thus, the energy can be written as:

$$E = \hat{U} + \frac{1}{2} m \vec{V} \cdot \vec{V}. \tag{5.9}$$

The physical origin of the internal energy stems from the internal kinetic and interatomic potential energies of the atoms making up the system. In other words, it is the energy of the internal "guts" of the system. If the system's mass does not change, then we can relate the energy to its specific energy via:

$$E = me. \tag{5.10}$$

Similarly, we can write the specific energy (by dividing Equation 5.9 with m) as:

$$e = \hat{u} + \frac{1}{2} \vec{V} \cdot \vec{V} \tag{5.11}$$

where $\hat{u}$ is the specific internal energy and has units of joule per kilogram, $\frac{J}{kg}$. Then writing the energy in terms of specific internal energy gives us:

$$E = m \left(\hat{u} + \frac{1}{2} \vec{V} \cdot \vec{V} \right). \tag{5.12}$$

The specific internal energy is not necessarily a useful property to be dealing with since it is not an easily measurable quantity. However, relationships can be developed between the specific internal energy and more measurable properties (such as temperature, pressure, etc.). Let's see if we can relate specific internal

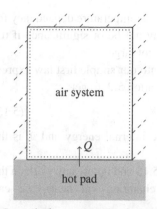
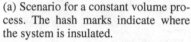
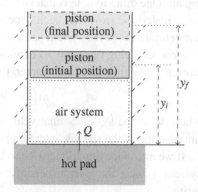

(a) Scenario for a constant volume process. The hash marks indicate where the system is insulated.

(b) Scenario for a constant pressure process. Note the piston is considered to be insulated.

Figure 5.1 Scenarios for discussing specific heat.

energy to something more tangible. To do so, consider a scenario given in 5.1a. In 5.1a a container of air is being heated from the bottom via a hot pad. All other sides are insulated (i.e., no "heat" is passing through the other sides). In this system, energy is being transferred to the system (from the surroundings) solely by heat. There is no work in this situation since there are no net forces acting on the system. Experimentally it turns out that for constant volume scenarios (such as in Figure 5.1a) we can relate the heat to the temperature change the system experiences via:

$$Q \propto m\left(T_f - T_i\right) \tag{5.13}$$

where T_i is the initial temperature of the system and T_f is the final temperature of the system. The m is again mass. We can add a proportionality constant, c_v, to the above equation to get:

$$Q = c_v m\left(T_f - T_i\right) \tag{5.14}$$

where c_v is called the specific heat, or more specifically in this case, the specific heat at constant volume. The units of specific heat are joules per kilogram per kelvin, or $\frac{J}{kgK}$, and the value is dependent on the material. The value of c_v under standard conditions for air is $416\frac{J}{kgK}$. For comparison, the c_v for water under similar conditions is roughly $4184\frac{J}{kgK}$. In general, the higher the specific heat value, the more energy it takes to raise or lower the temperature of the system per unit mass. Given that water has a much higher specific heat than air, the temperature of the oceans is not as variable as the temperature of the surrounding air. One thing to note is that specific heat has a temperature dependency for most materials. However, such dependency may not be of significance if the temperature range the system passes through is not large.

Now let's plug Equation 5.12 (with $\vec{V} = 0$) into our simple first law expression in the beginning of the chapter, that is, Equation 5.2, to get:

$$m\left(\hat{u}_f - \hat{u}_i\right) = Q - W \qquad (5.15)$$

where $\hat{u}_f$ is the system's final value of specific internal energy and $\hat{u}_i$ is the initial value of the specific internal energy.

If we introduce Equation 5.14 into Equation 5.15 and ignore any work on the system (since the system energy is only changing via heat), we can write:

$$m\left(\hat{u}_f - \hat{u}_i\right) = c_v m\left(T_f - T_i\right) - \underbrace{W}_{\text{ignore}}$$

$$\therefore \frac{\hat{u}_f - \hat{u}_i}{T_f - T_i} = c_v \qquad \text{(constant volume process)}. \qquad (5.16)$$

Equation 5.16 provides a simple relationship for the specific heat during a constant volume situation. It is equal to the change in specific internal energy divided by the change in temperature.

A more general definition of specific heat at constant volume is given by:

$$c_v = \left(\frac{\partial \hat{u}}{\partial T}\right)_v \qquad (5.17)$$

where the subscript v indicates the derivative is taken at a constant volume.

Now consider a second scenario given in Figure 5.1b. For this next scenario, the system is fitted with a piston (which is a movable lid). Suppose that as the system is heated from the bottom, the piston moves up such that the pressure in the container stays constant. If you measure the heat for this situation, you will find that it (again) is roughly proportional to the difference between the final and initial temperatures. However, the proportionality constant will be different than c_v. Instead, we will call this proportionality constant the specific heat at constant pressure (c_p):

$$Q \propto m\left(T_f - T_i\right)$$
$$\rightarrow Q = m c_p\left(T_f - T_i\right) \qquad \text{constant pressure process.} \qquad (5.18)$$

Keep in mind that in this situation, we have work as well as heat since there is a force needed to move the piston. If the piston is moving up (as shown in 5.1b), the air is doing work on the piston (i.e., it is pushing the piston up) and is thus transferring energy to the piston. On the other hand, if the piston is moving down (such as if the piston was too heavy), the air system would be having work done on it (i.e., energy would be transferred from the piston to the air). Work in this scenario is actually a relatively simple expression. You may recall that work can be loosely thought of as the force in the direction of motion multiplied by the distance traveled. The force in our case would just be the constant value of pressure (p) multiplied by the cross-sectional area of the piston (A). Therefore, the work is:

$$W = pA\left(y_f - y_i\right)$$

where $\left(y_f - y_i\right)$ is the distance traveled from an initial position, y_i to a final position, y_f. The A multiplied by the distance traveled is just a volume change, thus:

$$W = p\left(\mathcal{V}_f - \mathcal{V}_i\right) \quad \text{(constant pressure process)} \quad (5.19)$$

where $\mathcal{V}_i$ is the initial volume of air in the container and $\mathcal{V}_f$ is the final volume of air in the container. This work is often call $pd\mathcal{V}$ work.

We can write Equation 5.19 in terms of a specific volume ($\hat{v}$), which is just volume divided by mass, $\frac{\mathcal{V}}{m}$. Therefore we write volume as $m\hat{v}$ to make Equation 5.19 become:

$$W = pm\left(\hat{v}_f - \hat{v}_i\right). \quad (5.20)$$

We can plug Equations 5.20 and 5.18 into the Equation 5.15 to get:

$$m\left(\hat{u}_f - \hat{u}_i\right) = \underbrace{c_p m\left(T_f - T_i\right)}_{Q} - \underbrace{pm\left(\hat{v}_f - \hat{v}_i\right)}_{W}. \quad (5.21)$$

We can solve for c_p to get the following:

$$c_p = \frac{\left(\hat{u}_f + p\hat{v}_f\right) - \left(\hat{u}_i + p\hat{v}_i\right)}{T_f - T_i} \quad \text{(constant pressure process)}. \quad (5.22)$$

Notice that the specific heat at constant pressure appears to be related to the difference in the quantity, $\hat{u} + p\hat{v}$, divided by the difference in temperature. The $\hat{u} + p\hat{v}$ expression can be defined as a new thermodynamic property known as specific enthalpy, $\hat{h}$. The specific enthalpy can be written in two ways:

$$\hat{h} = \hat{u} + p\hat{v} \quad \text{or} \quad \hat{h} = \hat{u} + \frac{p}{\rho}. \quad (5.23)$$

Either definition is acceptable because the density (ρ) is nothing but the inverse of the specific volume ($\hat{v}$). With the inclusion of Equation 5.23 into Equation 5.22 we get:

$$c_p = \frac{\hat{h}_f - \hat{h}_i}{T_f - T_i}$$

or more formally if we take the limit as $T_f - T_i$ goes to zero:

$$c_p = \left(\frac{\partial \hat{h}}{\partial T}\right)_p \qquad (5.24)$$

where the subscript p denotes a constant pressure process.

You can think of specific heat (either at constant volume or constant pressure) as a material property since its value is mostly dependent on the material in question (as mentioned, there is also usually some temperature dependence as well). Of course, it is also considered a thermodynamic variable or property since it came directly from the application of the first law.

It turns out that it is very common in thermodynamics to define properties (usually dependent on the material) in terms of other thermodynamic variables. The specific heats are an example. Another example is the coefficient of thermal expansion, also called the bulk expansion coefficient. We will use the symbol β_T for the coefficient of thermal expansion. An expression for β_T is given by:

$$\beta_T = \frac{1}{\hat{v}}\left(\frac{\partial \hat{v}}{\partial T}\right)_p \quad \text{or} \quad \beta_T = -\frac{1}{\rho}\left(\frac{\partial \rho}{\partial T}\right)_p. \qquad (5.25)$$

The coefficient of thermal expansion is a measure of how much a material expands or contracts for given a temperature change. The units of β_T are inverse kelvin, or $\frac{1}{K}$. It is highly dependent on the material and hence we consider it a material property. Equation 5.25 calculates the change of volume (or density) with respect to a change in temperature at a given pressure (the subscript p indicates that the derivative is evaluated at constant pressure). The $\frac{1}{\hat{v}}$ (or $\frac{1}{\rho}$) is a scale factor to ensure that the specific volume (density) of the material is effectively "divided out." If the material expands when heated (which is typical), then the volume of the material should increase (or conversely, the density should decrease). Thus, the coefficient of thermal expansion is usually a positive quantity.

We will also come across another material property known as isothermal compressibility coefficient (β_p) as given by:

$$\beta_p = -\frac{1}{\hat{v}}\left(\frac{\partial \hat{v}}{\partial p}\right)_T \quad \text{or} \quad \beta_p = \frac{1}{\rho}\left(\frac{\partial \rho}{\partial p}\right)_T \qquad (5.26)$$

The units of this equation are inverse pressure, or $\frac{1}{Pa}$, where, as a reminder, Pa is the pascal given by $Pa = \frac{N}{m^2}$. This equation calculates the change of volume for a given pressure change (evaluated at a constant temperature, hence the subscript T in the derivative). The negative sign ensures a positive value for β_p because the volume of the material decreases with an increase in pressure. It can also be written in terms of density as shown in Equation 5.26.

You may have noticed that these thermodynamic derivatives are all evaluated assuming some variable being held constant (e.g., $\left(\frac{\partial \hat{h}}{\partial T}\right)_p$ is calculated at constant pressure, p). We can develop all kinds of various relationships relating thermodynamics properties to one another. One such relationship that will come in handy is the following:

$$\left(\frac{\partial \hat{u}}{\partial \hat{v}}\right)_T = \left(T\frac{\beta_T}{\beta_p} - p\right). \tag{5.27}$$

Another useful relationship for us (as you will see) is the change of enthalpy with respect to pressure at a constant temperature given by:

$$\left(\frac{\partial \hat{h}}{\partial p}\right)_T = (1 - T\beta_T)\,\hat{v}. \tag{5.28}$$

Developing Equations 5.27 and 5.28 requires a decent amount of thermodynamic formula gymnastics of which we will not show as it detracts from our purpose. However, these relationships will come in handy. At this point, we can move on to discussing the rate of heat and work.

5.1.2 Rate of Heat, $\dot{Q}$

Heat is considered to be the transfer of energy given a temperature difference.

In some ways, we can treat the rate of heat, $\dot{Q}$,[4] much like we treated mass flow rate ($\dot{m}$) in the first chapter in that we can write the rate of heat as an area integral of a flux quantity (i.e., heat flux, $\vec{q}''$) dotted with the normal to the surface. This would lead to:

$$\dot{Q} = - \oiint_{A(t)} \vec{q}'' \cdot \vec{n}dA. \tag{5.29}$$

A schematic of this scenario is given in Figure 5.2. Notice that we have a negative sign to ensure that the value for the rate of heat going into the fluid element is positive and negative if leaving.

An expression for the heat flux, $\vec{q}''$ was proposed in 1820 by Jean-Baptiste Joseph Fourier. Fourier proposed that the heat flux is proportional to a temperature difference with respect to distance. In one dimension, let's say the x-direction, Fourier's proposal would lead to the following expression:

$$q_x'' \propto \frac{\partial T}{\partial x}. \tag{5.30}$$

---

[4] Note, when we say the rate of heat, we actually mean rate of energy by the mechanism of heat. Heat is not a property that can be moved around and has a rate per se. It is instead a process or mechanism for which energy is transferred.

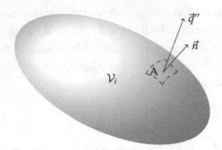

Figure 5.2 Illustration of heat flux.

where q_x'' is the heat flux in the x-direction.

Introducing a proportionality constant, k, known as the thermal conductivity, Equation 5.30 becomes:

$$q_x'' = -k\frac{\partial T}{\partial x}. \tag{5.31}$$

You might wonder why a negative sign was included in Equation 5.31. The reason stems from the second law of thermodynamics. The **second law of thermodynamics suggests heat can only spontaneously transfer from a hot temperature to a low temperature**. The negative sign ensures that this will be the case. To illustrate this, consider the situation given in Figure 5.3. In this scenario, we have a rectangle system (a block of iron for example) whose left end temperature is held fixed at 0°C and the right end is held fixed at 100°C. The length of the rectangle is 1 meter and all other sides of the rod are covered in rubber so that no heat "leaks" out the top and bottom sides. The heat flux (in the x-direction) at steady state can be approximated in the following manner:

$$q_x'' = -k\frac{\partial T}{\partial x} = -k\frac{\text{change in temperature}}{\text{change in x}} \approx -k\frac{0-100}{0-1} = -100k.$$

Notice that the value of the heat flux is $-100k$. Thus, the heat flux will be in the negative x-direction (shown in Figure 5.3 with the arrow) as long as the thermal conductivity is positive. In fact, the thermal conductivity needs to be positive in order to satisfy the second law of thermodynamics.

The thermal conductivity is a material property and thus its value depends on the material in question. The higher its value, the larger the heat flux will be for a given temperature gradient (i.e., temperature derivative). We say that high thermal conductivity materials are good thermal conductors and materials with low thermal conductivities are considered thermal insulators. The unit of thermal conductivity is a watt per meter per kelvin $\left(\frac{W}{mK}\right)$. Values of thermal

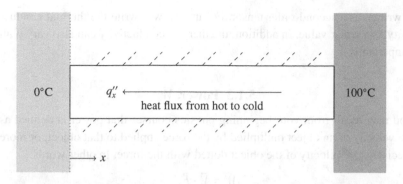

Figure 5.3 Schematic used to show the direction of heat is from hot to cold.

conductivity for pure copper at 20°C is 400 $\frac{W}{mK}$, water is roughly 0.6 $\frac{W}{mK}$, and air is 0.026 $\frac{W}{mK}$.

We can generalize the heat flux expression to multiple dimensions via:

$$\vec{q}'' = -k\vec{\nabla}T. \tag{5.32}$$

The heat flux in multiple dimensions is now just related to the gradient of temperature $(\vec{\nabla}T)$ instead of a single derivative $\left(\frac{\partial T}{\partial x}\right)$. In Cartesian coordinates for three dimensions, we have:

$$\vec{q}'' = -k\left(\hat{i}\frac{\partial}{\partial x} + \hat{j}\frac{\partial}{\partial y} + \hat{k}\frac{\partial}{\partial z}\right)T = -k\left(\hat{i}\frac{\partial T}{\partial x} + \hat{j}\frac{\partial T}{\partial y} + \hat{k}\frac{\partial T}{\partial z}\right).$$

Inserting Equation 5.32 into Equation 5.29 we get the following expression for the heat rate:

$$\dot{Q} = \oiint_{A(t)} \left(k\vec{\nabla}T\right) \cdot \vec{n}dA \tag{5.33}$$

We can convert Equation 5.33 into a volume integral using the divergence theorem to get:

$$\dot{Q} = \iiint_{V(t)} \vec{\nabla} \cdot \left(k\vec{\nabla}T\right)dV \tag{5.34}$$

We will use both Equation 5.33 and Equation 5.34 as we go forward.

Incidentally, the thermal conductivity can technically vary depending on direction. Such a material is called an anisotropic material. However, we are not going to consider such a scenario since, in fluid mechanics at least, this is not the typical scenario. When dealing with solid materials, however, this may become important. In the case of anisotropic materials, the thermal conductivity is written as a second-order tensor. For us, we will write the thermal conductivity as a scalar value. In addition, the thermal conductivity can also vary with temperature.

5.1.3 Power, $\dot{W}$

You may recall from your beginning physics courses that power is defined as the velocity of an object multiplied by the force applied to that object, or more precisely, the velocity of the object dotted with the force. In other words:

$$\dot{W} = \vec{V} \cdot \vec{F}.$$

In our case, the force on our fluid element can be written in two different ways. We obtained expressions for the force on a fluid in Chapter 3 when we discussed the stress tensor. We will take advantage of the area integral version of the total force acting on a fluid element, that is, Equation 3.8 which is given below as a reminder:

$$\Sigma\vec{F} = \oiint_{A(t)} \vec{\vec{T}} \cdot \vec{n} dA + \iiint_{V(t)} \rho\vec{g} dV \tag{3.8}$$

We are going to "sneak" the velocity vector into each integral and dot it with the integrand to get a power term:

$$\dot{W} = \underbrace{\oiint_{A(t)} \vec{V} \cdot \vec{\vec{T}} \cdot \vec{n} dA + \iiint_{V(t)} \rho\vec{V} \cdot \vec{g} dV}_{\text{not quite right for us yet}}.$$

There is one caveat with the expression we have obtained here though, our power term will actually be defined as the negative of what is given above. The reason for this is because of the sign convention we are using for power. We are using a sign convention where work (or power) that is done ON the system is negative. Thus, if you have a generic object (which does not have to be a fluid element) that is getting pulled to the right by some force (by let's say a string pulling on it), the work is done ON the object and should have a negative value. However, in the current equation for work (power), the work on the object being pulled by a string would be positive since the force and the velocity are in the same direction and thus their dot product would be positive. So, in order to be consistent with our sign convention, we need a negative sign in front of the power equation above. Thus, our equation for the power on a fluid element is given by:

$$\dot{W} = -\left(\oiint_{A(t)} \vec{V} \cdot \vec{\vec{T}} \cdot \vec{n} dA + \iiint_{V(t)} \rho\vec{V} \cdot \vec{g} dV \right) \tag{5.35}$$

Note that the last term of our power equation, the $\rho\vec{V}\cdot\vec{g}$ term, is considered to be a gravitational potential energy term.

We can use the divergence theorem on the area integral in Equation 5.35 to convert to a volume integral, just as we have done a number of times. Doing so allows us to write power as one volume integral:

$$\dot{W} = -\iiint_{\mathcal{V}(t)} \left(\vec{\nabla}\cdot\left(\vec{V}\cdot\vec{\bar{\tau}}\right) + \rho\vec{V}\cdot\vec{g} \right) d\mathcal{V} \qquad (5.36)$$

5.1.4 Source term

The final term to discuss in Equation 5.8 is the *Source* term. The *Source* term is included in situations where energy might be considered "generated" within the control volume. A very simple scenario to give you an idea of such an energy source might be to think about an electrical wire passing through your control volume as shown in the Cartesian control volume given in Figure 5.4. The electrical wire undoubtedly has a resistance and this resistance will cause the wire temperature to increase as electricity passes through it. A *Source* term might be useful to model the energy generated within the control volume that stems from the heat of the wire. In addition, the absorption of solar radiation within a system might be modeled as a *Source* term. The *Source* term can be written as a volume integral of an energy generation term on a per volume basis $(\dot{q}_{gen})$:

$$Source = \iiint_{\mathcal{V}} \dot{q}_{gen} d\mathcal{V}. \qquad (5.37)$$

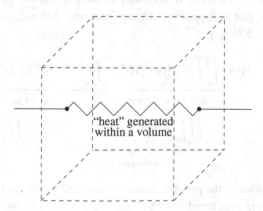

"heat" generated within a volume

Figure 5.4 Picture of wire passing through a Cartesian volume, which can be modeled as *Source* term via volumetric heat generation, $\dot{q}_{gen}$.

We call $\dot{q}_{gen}$ volumetric heat generation (or volumetric energy generation) and has units of $\frac{W}{m^3}$.

5.1.5 Putting it All Together

We can combine Equations 5.35, 5.33, and 5.37 with Equation 5.8 to get the **integral form of the energy equation** (note the stress tensor for a Newtonian fluid is also provided):

$$\iiint_V \frac{\partial(\rho e)}{\partial t}dV + \underbrace{\oiint_A (\rho e \vec{V}) \cdot \vec{n}dA}_{\text{flux of energy from flow}} = \underbrace{\oiint_A (k\vec{\nabla}T) \cdot \vec{n}dA}_{\dot{Q}}$$

$$-\underbrace{\left(-\left(\oiint_A \vec{V} \cdot \vec{\vec{T}} \cdot \vec{n}dA + \iiint_V \rho \vec{V} \cdot \vec{g}dV\right)\right)}_{\dot{W}} + \underbrace{\iiint_V \dot{q}_{gen}dV}_{Source}$$

Where:

$$e = \hat{u} + \frac{1}{2}\vec{V} \cdot \vec{V}$$

$$\vec{\vec{T}} = -p\vec{I} - \frac{2}{3}\mu\left(\vec{\nabla} \cdot \vec{V}\right)\vec{I} + \mu\left(\vec{\nabla}\vec{V} + \left(\vec{\nabla}\vec{V}\right)^{\dagger}\right)$$

$$(5.38)$$

To obtain a conservation form of the energy equation, we can convert the energy flux term (i.e., the second term on the left-hand side) to a volume integral using the divergence theorem. In addition, we need to use the volume integral forms for power, that is, Equation 5.36, and heat rate, that is, Equation 5.34, to modify Equation 5.38 to get:

$$\iiint_V \frac{\partial(\rho e)}{\partial t}dV + \underbrace{\iiint_V \vec{\nabla} \cdot (\rho e\vec{V})dV}_{\text{used divergence theorem}} = \underbrace{\iiint_V \vec{\nabla} \cdot (k\vec{\nabla}T)dV}_{\dot{Q} \text{ (used Equation 5.34)}}$$

$$-\underbrace{\left(-\iiint_V \left(\vec{\nabla} \cdot \left(\vec{V} \cdot \vec{\vec{T}}\right) + \rho\vec{V} \cdot \vec{g}\right)dV\right)}_{\dot{W} \text{ (used Equation 5.36)}} + \underbrace{\iiint_V \dot{q}_{gen}dV}_{Source}$$

As we have done in the previous sections, since all terms are now a volume integral integrating over an arbitrary volume, we can "shrink" the volume down to an infinitesimal size and effectively get rid of the integral. This leads to the **conservation form of the energy equation**:

$$\frac{\partial(\rho e)}{\partial t} + \underbrace{\vec{\nabla} \cdot (\rho e \vec{V})}_{\text{flux term}} = \underbrace{\vec{\nabla} \cdot (k\vec{\nabla}T)}_{\dot{Q}} - \underbrace{\left(-\left(\vec{\nabla} \cdot (\vec{V} \cdot \vec{\vec{T}}) + \rho \vec{V} \cdot \vec{g}\right)\right)}_{\dot{W}} + \underbrace{\dot{q}_{gen}}_{\text{Source}}$$

Where:

$$e = \hat{u} + \frac{1}{2}\vec{V} \cdot \vec{V}$$

$$\vec{\vec{T}} = -p\vec{\vec{I}} - \frac{2}{3}\mu(\vec{\nabla} \cdot \vec{V})\vec{\vec{I}} + \mu\left(\vec{\nabla}\vec{V} + (\vec{\nabla}\vec{V})^{\dagger}\right)$$

(5.39)

The left side of the conservation form of the equation can be re-written as:

$$\frac{\partial(\rho e)}{\partial t} + \vec{\nabla} \cdot (\rho e \vec{V}) = \underbrace{\rho\frac{\partial e}{\partial t} + e\frac{\partial \rho}{\partial t}}_{\substack{\text{product rule} \\ \text{applied to the} \\ \text{time derivative}}} + \underbrace{\rho e\vec{\nabla} \cdot \vec{V} + \rho\vec{V} \cdot \vec{\nabla}e + \overbrace{e\vec{V} \cdot \vec{\nabla}\rho}^{\vec{V} \cdot \vec{\nabla}(\rho e)}}_{\substack{\text{the flux term "broken" up} \\ \text{using a process similar} \\ \text{to the product rule}}}.$$

The last three terms came from breaking out (using some vector calculus rules) the $\vec{\nabla} \cdot (\rho e \vec{V})$ term. We can rearrange the right hand side of the above equation to get:

$$\frac{\partial(\rho e)}{\partial t} + \vec{\nabla} \cdot (\rho e \vec{V}) = \rho\left(\frac{\partial e}{\partial t} + \vec{V} \cdot \vec{\nabla}e\right) + e\underbrace{\left(\frac{\partial \rho}{\partial t} + \rho\vec{\nabla} \cdot \vec{V} + \vec{V} \cdot \vec{\nabla}\rho\right)}_{\text{non-conservation form of continuity}}$$

Notice that the last three terms in the above equation look like the non-conservation form of the continuity equation (which is equal to zero). Thus, the above equation becomes:

$$\frac{\partial(\rho e)}{\partial t} + \vec{\nabla} \cdot (\rho e \vec{V}) = \rho\left(\frac{\partial e}{\partial t} + \vec{V} \cdot \vec{\nabla}e\right).$$

(5.40)

The right hand side of Equation 5.40 contains a material derivative of specific energy (in an Eulerian description), which is indicative of the non-conservation form. Thus, plugging the relationship given in Equation 5.40 into Equation 5.39, we get the **non-conservation form of the energy equation**:

$$\rho\underbrace{\left(\frac{\partial e}{\partial t} + \vec{V} \cdot \vec{\nabla}e\right)}_{\frac{De}{Dt}} = \underbrace{\vec{\nabla} \cdot \left(k\vec{\nabla}T\right)}_{\dot{Q}} - \underbrace{\left(-\left(\vec{\nabla} \cdot \left(\vec{V} \cdot \vec{\vec{T}}\right) + \rho\vec{V} \cdot \vec{g}\right)\right)}_{\dot{W}} + \underbrace{\dot{q}_{gen}}_{Source}$$

Where:

$$e = \hat{u} + \frac{1}{2}\vec{V} \cdot \vec{V}$$

$$\vec{\vec{T}} = -p\vec{\vec{I}} - \frac{2}{3}\mu\left(\vec{\nabla} \cdot \vec{V}\right)\vec{\vec{I}} + \mu\left(\vec{\nabla}\vec{V} + \left(\vec{\nabla}\vec{V}\right)^{\dagger}\right)$$

(5.41)

Finally, recognizing that the non-conservation form is easily converted to Lagrangian form by replacing the material derivative of specific energy on the left-hand side of Equation 5.41 with the Lagrangian form for the material derivative, that is, using $\frac{De}{Dt}$. Thus, we can write the **Lagrangian form of the energy equation** as:

$$\rho\frac{De}{Dt} = \underbrace{\vec{\nabla} \cdot \left(k\vec{\nabla}T\right)}_{\dot{Q}} - \underbrace{\left(-\left(\vec{\nabla} \cdot \left(\vec{V} \cdot \vec{\vec{T}}\right) + \rho\vec{V} \cdot \vec{g}\right)\right)}_{\dot{W}} + \underbrace{\dot{q}_{gen}}_{Source}$$

Where:

$$e = \hat{u} + \frac{1}{2}\vec{V} \cdot \vec{V}$$

$$\vec{\vec{T}} = -p\vec{\vec{I}} - \frac{2}{3}\mu\left(\vec{\nabla} \cdot \vec{V}\right)\vec{\vec{I}} + \mu\left(\vec{\nabla}\vec{V} + \left(\vec{\nabla}\vec{V}\right)^{\dagger}\right)$$

(5.42)

5.2 A Very Common Form of the Energy Equation

We can plug Equation 5.11 ($e = \hat{u} + \frac{1}{2}\vec{V} \cdot \vec{V}$) into the general Lagrangian form of the energy equation (Equation 5.42) to get (with the negative sign on the work term distributed through the parentheses):

$$\rho\frac{D\hat{u}}{Dt} + \frac{\rho}{2}\frac{D\left(\vec{V} \cdot \vec{V}\right)}{Dt} = \vec{\nabla} \cdot \left(k\vec{\nabla}T\right) + \vec{\nabla} \cdot \left(\vec{V} \cdot \vec{\vec{T}}\right) + \rho\vec{V} \cdot \vec{g} + \dot{q}_{gen}. \quad (5.43)$$

Let's take a closer look at the $\vec{\nabla} \cdot \left(\vec{V} \cdot \vec{\vec{T}}\right)$ term. We are going to distribute the ∇ symbol through the parentheses by performing an operation very much like the product rule in calculus to get:

$$\vec{\nabla} \cdot \left(\vec{V} \cdot \vec{\vec{T}}\right) = \vec{V} \cdot \left(\vec{\nabla} \cdot \vec{\vec{T}}\right) + \vec{\vec{T}} : \vec{\nabla}\vec{V}. \quad (5.44)$$

The last term on the right side of the equal sign contains a new operation, the double dot product (i.e., the : operation). The double dot product operation is a second-order tensor operation. In many ways, you can think of it as an operation that is similar to a dot product for vectors except now generalized to second-order tensors, where the second order tensors in this case are $\vec{\vec{T}}$ and $\vec{\nabla}\vec{V}$. In Cartesian coordinates, this amounts to nothing but taking each element of $\vec{\vec{T}}$ and multiplying it with the corresponding element of $\vec{\nabla}\vec{V}$ and adding them up. Observe, using Equation 3.5 for the stress tensor in Cartesian coordinates and Equation 3.20 for the velocity gradient in Cartesian coordinates:

$$\vec{\vec{T}} : \vec{\nabla}\vec{V} = \underbrace{\begin{pmatrix} \tau_{xx} & \tau_{xy} & \tau_{xz} \\ \tau_{yx} & \tau_{yy} & \tau_{yz} \\ \tau_{zx} & \tau_{zy} & \tau_{zz} \end{pmatrix}}_{\text{Equation 3.5}} : \underbrace{\begin{pmatrix} \frac{\partial u}{\partial x} & \frac{\partial v}{\partial x} & \frac{\partial w}{\partial x} \\ \frac{\partial u}{\partial y} & \frac{\partial v}{\partial y} & \frac{\partial w}{\partial y} \\ \frac{\partial u}{\partial z} & \frac{\partial v}{\partial z} & \frac{\partial w}{\partial z} \end{pmatrix}}_{\text{Equation 3.20}}$$

$$= \tau_{xx}\frac{\partial u}{\partial x} + \tau_{xy}\frac{\partial v}{\partial x} + \tau_{xz}\frac{\partial w}{\partial x}$$
$$+ \tau_{yx}\frac{\partial u}{\partial y} + \tau_{yy}\frac{\partial v}{\partial y} + \tau_{yz}\frac{\partial w}{\partial y}$$
$$+ \tau_{zx}\frac{\partial u}{\partial z} + \tau_{zy}\frac{\partial v}{\partial z} + \tau_{zz}\frac{\partial w}{\partial z}.$$

This term will contain something known as viscous dissipation, which we will discuss later. For now, just think of the double dot product as an operation that is similar to a dot product for vectors except now generalized to second-order tensors.

With the introduction of Equation 5.44 into Equation 5.43 we get:

$$\rho\frac{D\hat{u}}{Dt} + \frac{\rho}{2}\frac{D(\vec{V}\cdot\vec{V})}{Dt} = \vec{\nabla}\cdot\left(k\vec{\nabla}T\right) + \vec{V}\cdot\left(\vec{\nabla}\cdot\vec{\vec{T}}\right) + \vec{\vec{T}} : \vec{\nabla}\vec{V} + \rho\vec{V}\cdot\vec{g} + \dot{q}_{gen}.$$
$$(5.45)$$

Now consider that the kinetic energy term $\left(\frac{\rho}{2}\frac{D(\vec{V}\cdot\vec{V})}{Dt}\right)$ can be expanded out through the product rule of calculus via:

$$\frac{\rho}{2}\frac{D(\vec{V}\cdot\vec{V})}{Dt} = \frac{\rho}{2}\underbrace{\left(\vec{V}\cdot\frac{D\vec{V}}{Dt} + \vec{V}\cdot\frac{D\vec{V}}{Dt}\right)}_{2\vec{V}\cdot\frac{D\vec{V}}{Dt}} = \rho\vec{V}\cdot\frac{D\vec{V}}{Dt}. \qquad (5.46)$$

We can plug Equation 5.46 into Equation 5.45 to get (note, we numerically label some terms because they will be useful):

$$\rho\frac{D\hat{u}}{Dt} + \underbrace{\rho\vec{V}\cdot\frac{D\vec{V}}{Dt}}_{1} = \vec{\nabla}\cdot\left(k\vec{\nabla}T\right) + \underbrace{\vec{V}\cdot\left(\vec{\nabla}\cdot\vec{\overline{T}}\right)+\vec{\overline{T}}:\vec{\nabla}\vec{V}}_{2} + \underbrace{\rho\vec{V}\cdot\vec{g}}_{3}+\dot{q}_{gen}. \quad (5.47)$$

Notice something very useful. Equation 5.47 has terms marked 1, 2, and 3. These terms together are nothing but Cauchy's first law discussed in Chapter 3 dotted with the velocity vector. As a reminder, Cauchy's first law in Langrangian form is Equation 3.11 from Chapter 3, re-written here:

$$\rho\frac{D\vec{V}}{Dt} = \vec{\nabla}\cdot\vec{\overline{T}} + \rho\vec{g}.$$

If we dot Cauchy's first law with the velocity vector, we get:

$$\vec{V}\cdot\left(\rho\frac{D\vec{V}}{Dt}\right) = \vec{V}\cdot\left(\vec{\nabla}\cdot\vec{\overline{T}}\right) + \vec{V}\cdot\left(\rho\vec{g}\right)$$

$$\rightarrow \underbrace{\rho\vec{V}\cdot\frac{D\vec{V}}{Dt}}_{\text{matches 1}} = \underbrace{\vec{V}\cdot\left(\vec{\nabla}\cdot\vec{\overline{T}}\right)}_{\text{matches 2}} + \underbrace{\rho\vec{V}\cdot\vec{g}}_{\text{matches 3}}. \quad (5.48)$$

The terms from Equation 5.48 match the 1, 2, and 3 terms from Equation 5.47. Equation 5.48 is a variation of what is called the **mechanical energy equation**. We can substitute Equation 5.48 into Equation 5.47 to get a much simpler form (effectively eliminating the 1, 2, and 3 terms):

$$\rho\frac{D\hat{u}}{Dt} = \vec{\nabla}\cdot\left(k\vec{\nabla}T\right) + \vec{\overline{T}}:\vec{\nabla}\vec{V} + \dot{q}_{gen}. \quad (5.49)$$

Equation 5.49 is a general energy equation with Cauchy's momentum equation "factored" out. Equation 5.49 is considered a **thermal energy equation**, however, it is often still just called the **energy equation**. It can apply to a fluid or a solid depending on the form of the stress tensor. The form of the stress tensor we are using is the Newtonian fluid stress tensor, which is (again):

$$\vec{\overline{T}} = -p\vec{\overline{I}} - \frac{2}{3}\mu\left(\vec{\nabla}\cdot\vec{V}\right)\vec{\overline{I}} + \mu\left(\vec{\nabla}\vec{V} + \left(\vec{\nabla}\vec{V}\right)^{\dagger}\right).$$

Including the stress tensor into Equation 5.49 gives us:

$$\rho\frac{D\hat{u}}{Dt} = \vec{\nabla}\cdot\left(k\vec{\nabla}T\right)+\left(-p\vec{\overline{I}} - \frac{2}{3}\mu\left(\vec{\nabla}\cdot\vec{V}\right)\vec{\overline{I}} + \mu\left(\vec{\nabla}\vec{V} + \left(\vec{\nabla}\vec{V}\right)^{\dagger}\right)\right):\vec{\nabla}\vec{V}+\dot{q}_{gen}. \quad (5.50)$$

Distributing the double dot product through to each term yields:

$$\rho\frac{D\hat{u}}{Dt} = \vec{\nabla}\cdot\left(k\vec{\nabla}T\right) - p\vec{I}: \vec{\nabla}\vec{V} - \frac{2}{3}\mu\left(\vec{\nabla}\cdot\vec{V}\right)\vec{I}: \vec{\nabla}\vec{V} + \mu\left(\vec{\nabla}\vec{V} + \left(\vec{\nabla}\vec{V}\right)^{\dagger}\right): \vec{\nabla}\vec{V} + \dot{q}_{gen}. \quad (5.51)$$

We can make some more simplifications to the double dot product terms for pressure (i.e., $-p\vec{I}: \vec{\nabla}\vec{V}$) and the divergence of velocity term (i.e., $-\frac{2}{3}\mu\left(\vec{\nabla}\cdot\vec{V}\right)\vec{I}: \vec{\nabla}\vec{V}$) in Equation 5.51. Let's start with the double dot product for pressure:

$$-p\vec{I}: \vec{\nabla}\vec{V} = \begin{pmatrix} -p & 0 & 0 \\ 0 & -p & 0 \\ 0 & 0 & -p \end{pmatrix} : \begin{pmatrix} \frac{\partial u}{\partial x} & \frac{\partial v}{\partial x} & \frac{\partial w}{\partial x} \\ \frac{\partial u}{\partial y} & \frac{\partial v}{\partial y} & \frac{\partial w}{\partial y} \\ \frac{\partial u}{\partial z} & \frac{\partial v}{\partial z} & \frac{\partial w}{\partial z} \end{pmatrix}$$

$$= -p\frac{\partial u}{\partial x} + 0\frac{\partial v}{\partial x} + 0\frac{\partial w}{\partial x}$$
$$+ 0\frac{\partial u}{\partial y} - p\frac{\partial v}{\partial y} + 0\frac{\partial w}{\partial y} \qquad (5.52)$$
$$+ 0\frac{\partial u}{\partial z} + 0\frac{\partial v}{\partial z} - p\frac{\partial w}{\partial z}$$

$$= -p\underbrace{\left(\frac{\partial u}{\partial x} + \frac{\partial v}{\partial y} + \frac{\partial w}{\partial z}\right)}_{\vec{\nabla}\cdot\vec{V}}.$$

Given the result of breaking out the pressure term double dotted with the velocity gradient, we can conclude:

$$-p\vec{I}: \vec{\nabla}\vec{V} = -p\vec{\nabla}\cdot\vec{V}. \quad (5.53)$$

We can follow a similar procedure for the $-\frac{2}{3}\mu\left(\vec{\nabla}\cdot\vec{V}\right)\vec{I}: \vec{\nabla}\vec{V}$ term to get:

$$-\frac{2}{3}\mu\left(\vec{\nabla}\cdot\vec{V}\right)\vec{I}: \vec{\nabla}\vec{V} = -\frac{2}{3}\mu\left(\vec{\nabla}\cdot\vec{V}\right)^{2}. \quad (5.54)$$

Using Equation 5.53 and Equation 5.54 in Equation 5.51 we get the **"thermal" energy equation in Lagrangian form for a Newtonian fluid**:

$$\boxed{\rho\frac{D\hat{u}}{Dt} = \vec{\nabla}\cdot\left(k\vec{\nabla}T\right) - p\vec{\nabla}\cdot\vec{V} - \frac{2}{3}\mu\left(\vec{\nabla}\cdot\vec{V}\right)^{2} + \mu\left(\vec{\nabla}\vec{V} + \left(\vec{\nabla}\vec{V}\right)^{\dagger}\right): \vec{\nabla}\vec{V} + \dot{q}_{gen}}$$
$$(5.55)$$

With Equation 5.55 we are close to having a complete energy equation for a Newtonian fluid that will be useful for our needs. However, we currently have specific internal energy, $\hat{u}$, on the left-hand side of the equal sign. It would be nice if we could convert $\hat{u}$ into more useful (and more readily observable) thermodynamic properties such as temperature, density, and pressure. It turns out that we can do this thanks to a thermodynamic principle known as the state principle. The **state principle** in thermodynamics is the idea that any intensive property (recall an intensive property does not depend on the size of the system) can be written in terms of two other intensive properties. As an example, suppose we knew the specific volume and temperature of a system (both of which are intensive properties), then we can obtain (in theory) any other intensive property such as specific internal energy, specific enthalpy, pressure, etc. The state principle is essentially an empirical observation that is given considerable weight in thermodynamics.

In our particular case, we have specific internal energy on the left-hand side of the equal sign of Equation 5.55. If we considered the internal energy to be a function of temperature and specific volume, that is, $\hat{u} = f(T, \hat{v})$, then we can rewrite the material derivative of the specific internal energy in terms of temperature and density using the chain rule from calculus:

$$\frac{D\hat{u}}{Dt} = \underbrace{\left(\frac{\partial \hat{u}}{\partial T}\right)_v}_{c_v} \frac{DT}{Dt} + \underbrace{\left(\frac{\partial \hat{u}}{\partial \hat{v}}\right)_T}_{T\frac{\beta_T}{\beta_p} - p} \frac{D\hat{v}}{Dt}. \tag{5.56}$$

Recalling from Equations 5.17 and 5.27 that $\left(\frac{\partial \hat{u}}{\partial T}\right)_v$ is the specific heat at constant volume (c_v) and $\left(\frac{\partial \hat{u}}{\partial \hat{v}}\right)_T$ is $\left(T\frac{\beta_T}{\beta_p} - p\right)$, respectively, Equation 5.56 becomes:

$$\frac{D\hat{u}}{Dt} = c_v \frac{DT}{Dt} + \left(T\frac{\beta_T}{\beta_p} - p\right)\frac{D\hat{v}}{Dt}.$$

The material derivative of specific volume is given by the same relationship as the material derivative of an infinitesimal volume as was discussed in Chapter 2 (Equation 2.33), that is:

$$\frac{1}{\hat{v}}\frac{D\hat{v}}{Dt} = \vec{\nabla} \cdot \vec{V}. \tag{5.57}$$

Plugging Equation 5.57 into Equation 5.56 gives:

$$\frac{D\hat{u}}{Dt} = c_v \frac{DT}{Dt} + \left(T\frac{\beta_T}{\beta_p} - p\right)\hat{v}\vec{\nabla} \cdot \vec{V}. \tag{5.58}$$

In fluid mechanics, we typically deal with density instead of specific volume, and density is just the inverse of specific volume ($\rho = \frac{1}{\hat{v}}$), thus Equation 5.58 becomes:

$$\boxed{\frac{D\hat{u}}{Dt} = c_v \frac{DT}{Dt} + \frac{1}{\rho}\left(T\frac{\beta_T}{\beta_p} - p\right)\vec{\nabla}\cdot\vec{V}} \qquad (5.59)$$

Equation 5.59 will work for the material derivative of specific internal energy. However, for various (albeit subtle) reasons, the material derivative of specific internal energy can also be written differently by taking advantage of the definition of specific enthalpy:

$$\hat{h} = \hat{u} + p\hat{v}.$$

Using the above equation for specific enthalpy, the material derivative of the specific internal energy leads to:

$$\frac{D\hat{u}}{Dt} = \frac{D}{Dt}\left(\hat{h} - p\hat{v}\right)$$

$$= \frac{D\hat{h}}{Dt} - \frac{D}{Dt}(p\hat{v})$$

$$= \frac{D\hat{h}}{Dt} - \hat{v}\frac{Dp}{Dt} - p\frac{D\hat{v}}{Dt}.$$

Using Equation 5.57 and the fact that $\hat{v} = \frac{1}{\rho}$, we get:

$$\frac{D\hat{u}}{Dt} = \frac{D\hat{h}}{Dt} - \frac{1}{\rho}\frac{Dp}{Dt} - \frac{p}{\rho}\vec{\nabla}\cdot\vec{V}. \qquad (5.60)$$

We can now use the chain rule to expand out the material derivative of specific enthalpy in terms of more meaningful variables. For example, if we assume that the specific enthalpy is written as a function of temperature and pressure, $h = f(p, T)$, then we can write (using calculus again):

$$\frac{D\hat{h}}{Dt} = \underbrace{\left(\frac{\partial\hat{h}}{\partial T}\right)_p}_{c_p}\frac{DT}{Dt} + \underbrace{\left(\frac{\partial\hat{h}}{\partial p}\right)_T}_{(1-\beta_T T)\hat{v}}\frac{Dp}{Dt} \qquad (5.61)$$

$$\rightarrow \frac{D\hat{h}}{Dt} = c_p\frac{DT}{Dt} + (1 - \beta_T T)\,\hat{v}\frac{Dp}{Dt}.$$

We took advantage of Equations 5.24 and 5.28 to get Equation 5.61. Plugging in Equation 5.61 for the material derivative of specific enthalpy into Equation 5.60 and replacing the $\hat{v}$ with an inverse of density ($\frac{1}{\rho}$), the equation for the material derivative of specific internal energy becomes:

$$\frac{D\hat{u}}{Dt} = c_p\frac{DT}{Dt} + \frac{1 - \beta_T T}{\rho}\frac{Dp}{Dt} - \frac{1}{\rho}\frac{Dp}{Dt} - \frac{p}{\rho}\vec{\nabla}\cdot\vec{V}$$

leading to:

$$\frac{D\hat{u}}{Dt} = c_p \frac{DT}{Dt} - \frac{\beta_T T}{\rho} \frac{Dp}{Dt} - \frac{p}{\rho} \vec{\nabla} \cdot \vec{V} \qquad (5.62)$$

We can use either 5.59 or Equation 5.62 for the material derivative of the specific internal energy. Plugging these equations into Equation 5.55 yield:

$$\rho \left(c_v \frac{DT}{Dt} + \frac{1}{\rho} \left(T\frac{\beta_T}{\beta_p} - p \right) \vec{\nabla} \cdot \vec{V} \right) = \vec{\nabla} \cdot \left(k\vec{\nabla}T \right) - p\vec{\nabla} \cdot \vec{V} - \frac{2}{3}\mu \left(\vec{\nabla} \cdot \vec{V} \right)^2$$
$$+ \mu \left(\vec{\nabla}\vec{V} + \left(\vec{\nabla}\vec{V} \right)^\dagger \right) : \vec{\nabla}\vec{V} + \dot{q}_{gen}$$

$$(5.63)$$

and

$$\rho \left(c_p \frac{DT}{Dt} - \frac{\beta_T T}{\rho} \frac{Dp}{Dt} - \frac{p}{\rho} \vec{\nabla} \cdot \vec{V} \right) = \vec{\nabla} \cdot \left(k\vec{\nabla}T \right) - p\vec{\nabla} \cdot \vec{V} - \frac{2}{3}\mu \left(\vec{\nabla} \cdot \vec{V} \right)^2$$
$$+ \mu \left(\vec{\nabla}\vec{V} + \left(\vec{\nabla}\vec{V} \right)^\dagger \right) : \vec{\nabla}\vec{V} + \dot{q}_{gen}$$

$$(5.64)$$

Notice that, for both Equation 5.64 and Equation 5.63, that there is a $-p\vec{\nabla} \cdot \vec{V}$ term on both sides of the equal sign that we can eliminate. In addition, we can move all of the terms except for the material derivative of temperature to the right-hand side of the equations to get the **"thermal" energy equation (using c_v and c_p) for a Newtonian fluid in Lagrangian form**:

$$\rho c_v \frac{DT}{Dt} = \vec{\nabla} \cdot \left(k\vec{\nabla}T \right) - \frac{2}{3}\mu \left(\vec{\nabla} \cdot \vec{V} \right)^2$$
$$+ \mu \left(\vec{\nabla}\vec{V} + \left(\vec{\nabla}\vec{V} \right)^\dagger \right) : \vec{\nabla}\vec{V} - T\left(\frac{\beta_T}{\beta_p} \right) \vec{\nabla} \cdot \vec{V} + \dot{q}_{gen}$$

$$(5.65)$$

and

$$\rho c_p \frac{DT}{Dt} = \vec{\nabla} \cdot \left(k\vec{\nabla}T \right) - \frac{2}{3}\mu \left(\vec{\nabla} \cdot \vec{V} \right)^2$$
$$+ \mu \left(\vec{\nabla}\vec{V} + \left(\vec{\nabla}\vec{V} \right)^\dagger \right) : \vec{\nabla}\vec{V} + \beta_T T \frac{Dp}{Dt} + \dot{q}_{gen}$$

$$(5.66)$$

Notice Equation 5.65 contains a $\frac{\beta_T}{\beta_p}$ term, which can approach the unfortunate ratio of $\frac{0}{0}$ under certain circumstances, such as for a constant density situation. Thus, even though both of these versions of the energy equation are correct, Equation 5.66 is typically the preferred one and, as such, we will be using this version as we proceed. In addition, from here on out we will refer to Equation 5.66 as the energy equation as opposed to the thermal energy equation.

5.3 Initial Discussion of the Energy Equation

There were quite a few versions of the energy equation that we introduced. We are going to focus our attention on Equation 5.66 as well as its non-conservation variation. The non-conservation variation only requires that we break the material derivatives of temperature and pressure out into an Eulerian description. In addition, we will also spend a good deal of time on the incompressible flow version of the energy equation.

Like we did for the Navier–Stokes equations, we can label some terms for the energy equation. In particular, we can label some of the terms from Equation 5.66 with the material derivative of temperature expanded out:

Compressible Energy Equation

$$\rho c_p \underbrace{\left(\underbrace{\frac{\partial T}{\partial t}}_{\substack{\text{local time} \\ \text{derivative}}} + \underbrace{\vec{V} \cdot \vec{\nabla} T}_{\substack{\text{advective} \\ \text{transport}}} \right)}_{\substack{\text{material derivative} \\ \text{of temperature, } \frac{DT}{Dt}}} = \overset{\substack{\text{diffusive transport} \\ \text{(conduction)}}}{\overbrace{\vec{\nabla} \cdot \left(k \vec{\nabla} T \right)}} \; \overset{\substack{\text{work of} \\ \text{volume change}}}{\overbrace{- \frac{2}{3} \mu \left(\vec{\nabla} \cdot \vec{V} \right)^2}}$$

with the brace over the left side reading "change of energy within a fluid with respect to time"

$$+ \underbrace{\mu \left(\vec{\nabla} \vec{V} + \left(\vec{\nabla} \vec{V} \right)^\dagger \right) : \vec{\nabla} \vec{V}}_{\text{frictional heating}} + \underbrace{\beta_T T \frac{Dp}{Dt}}_{\substack{\text{energy due} \\ \text{to thermal} \\ \text{expansion}}} + \underbrace{\dot{q}_{gen}}_{\substack{\text{energy} \\ \text{generation} \\ \text{within fluid}}}$$

Additional variables: c_p - specific heat at constant pressure $\left(\frac{\text{J}}{\text{kg K}} \right)$

k - thermal conductivity $\left(\frac{\text{W}}{\text{m K}} \right)$

β_T - thermal expansion coefficient $\left(\frac{1}{\text{K}} \right)$

Let's discuss briefly each term:

$\rho c_p \left(\frac{\partial T}{\partial t} + \vec{V} \cdot \vec{\nabla} T \right)$: This is the whole left side of the energy equation and constitutes an energy storage term. The specific heat, c_p, is the material parameter dealing with how much energy is needed to raise the temperature of a material (or fluid) on a per mass basis. The higher the specific heat, the more energy is needed for a given temperature change. The density multiplied by the specific heat (ρc_p) makes the energy storage term on a per volume basis. The term in parentheses on the left-hand side (i.e., the material derivative of temperature) was studied in some detail in Chapter 2. There, we studied the material derivative of temperature quite a bit when it was set to zero. In the case of the energy equation, it no longer equals zero. Instead, it is equal to the transfer of energy due to heat, work, and energy generation. If we think in terms of a Lagrangian description, the temperature of a moving fluid element can change due to the heat passing through it (via diffusion), the work being done on or by it, and any energy generation within the fluid element. It will be most illustrative to see what happens when this term and the diffusion term are combined (with the others neglected). We will discuss this scenario a little later when we cover the convection–diffusion equation.

$\vec{\nabla} \cdot \left(k \vec{\nabla} T \right)$: This is called the diffusion or **diffusive transport** term. More specifically, the thermal diffusion term. It is also known as the **conduction** term. This term stems directly from the transfer of energy by heat via Fourier's law of conduction, Equation 5.32. This very important term will be discussed at length using two model partial differential equations: Laplace's equation and the heat equation.

$-\frac{2}{3}\mu \left(\vec{\nabla} \cdot \vec{V} \right)^2$ and $\beta_T T \frac{Dp}{Dt}$: These two terms are both related to the energy (or work) due to a volume change of a fluid element. From Chapter 2 you know that the divergence of velocity is related to the volume change of a fluid element. Thus, the divergence of velocity in $-\frac{2}{3}\mu \left(\vec{\nabla} \cdot \vec{V} \right)^2$ implies an energy transfer term related to the change in the fluid element volume. In addition, due to the isothermal expansion coefficient (β_T), the $\beta_T T \frac{Dp}{Dt}$ term also deals with volume change. Recall that the isothermal expansion coefficient is related to the change of volume of a fluid element for a given pressure change. For an incompressible fluid, the isothermal expansion coefficient is small and thus this term is usually neglected for incompressible flows.

$\mu \left(\vec{\nabla}\vec{V} + \left(\vec{\nabla}\vec{V} \right)^{\dagger} \right) : \vec{\nabla}\vec{V}$: This term is a frictional heating term. When combined with $-\frac{2}{3}\mu \left(\vec{\nabla} \cdot \vec{V} \right)^2$, it is called **viscous dissipation**. We will break out this term in Cartesian coordinates shortly and, in doing so, will illustrate that this term will always yield a positive value. Thus, since this term is positive, it only contributes to increasing (as opposed to decreasing) the temperature of a moving fluid. The reason behind this positive contribution is due to the fact that this term is essentially work due to shearing (or friction). Recall from

Chapter 3 that the gradient of velocity can usually be thought of as a friction type of force. Friction only works in one way, that is, friction work is always done on a system and thus will contribute to increasing the temperature through the dissipation of work into heat. Much like rubbing your hands together can only increase the temperature of your hands as opposed to decreasing the temperature of your hands. The same can be said for fluids undergoing shearing forces. This effect is related to the second law of thermodynamics. In words, friction is always present and in some sense can be thought of as "wasted" energy. Wasted in the form of causing an increase in temperature as opposed to using that energy to do some useful work.

$\dot{q}_{gen}$: The energy generation term was discussed earlier in the chapter. An example of when this term might be needed is a situation where you want to model thermal radiation being absorbed within the body.

5.3.1 Incompressible Flows

The energy equation given in Equation 5.66 can be simplified for incompressible flows by recognizing that the $\left(\vec{\nabla} \cdot \vec{V}\right)^2$ term will go away. In addition, for incompressible flows, the β_T term will go away as well. Thus, the energy equation can be simplified to get the **energy equation for incompressible flows in Lagrangian form**:

$$\rho c_p \frac{DT}{Dt} = \vec{\nabla} \cdot \left(k\vec{\nabla}T\right) + \mu\left(\vec{\nabla}\vec{V} + \left(\vec{\nabla}\vec{V}\right)^\dagger\right) : \vec{\nabla}\vec{V} + \dot{q}_{gen}. \quad (5.67)$$

Furthermore, if we assume a constant thermal conductivity, we can simplify the energy equation even more to get the **energy equation for incompressible flows in Lagrangian form with a constant thermal conductivity**:

$$\boxed{\rho c_p \frac{DT}{Dt} = k\nabla^2 T + \mu\left(\vec{\nabla}\vec{V} + \left(\vec{\nabla}\vec{V}\right)^\dagger\right) : \vec{\nabla}\vec{V} + \dot{q}_{gen}} \quad (5.68)$$

We can expand out the material derivative to get the **energy equation for incompressible flows in non-conservation form with a constant thermal conductivity**:

$$\boxed{\rho c_p \left(\frac{\partial T}{\partial t} + \vec{V} \cdot \vec{\nabla}T\right) = k\nabla^2 T + \mu\left(\vec{\nabla}\vec{V} + \left(\vec{\nabla}\vec{V}\right)^\dagger\right) : \vec{\nabla}\vec{V} + \dot{q}_{gen}} \quad (5.69)$$

Again, we can label some terms:

Incompressible Energy Equation (non-conservation form)

$$\rho c_p \underbrace{\left(\underbrace{\frac{\partial T}{\partial t}}_{\substack{\text{local time} \\ \text{derivative}}} + \underbrace{\vec{V} \cdot \vec{\nabla} T}_{\substack{\text{advective} \\ \text{transport}}} \right)}_{} = \underbrace{k\nabla^2 T}_{\substack{\text{diffusive} \\ \text{transport} \\ \text{(conduction)}}} + \underbrace{\mu \left(\vec{\nabla}\vec{V} + \left(\vec{\nabla}\vec{V}\right)^\dagger \right) : \vec{\nabla}\vec{V}}_{\substack{\text{viscous dissipation} \\ \text{(frictional heating)}}} + \dot{q}_{gen}$$

Note, we are now labeling the $\mu \left(\vec{\nabla}\vec{V} + \left(\vec{\nabla}\vec{V}\right)^\dagger \right) : \vec{\nabla}\vec{V}$ as viscous dissipation since there is no longer a $-\frac{2}{3}\mu \left(\vec{\nabla} \cdot \vec{V} \right)^2$ term.

We have already seen the two dimensional incompressible continuity and Navier–Stokes equations from Equations 4.24, 4.25, and 4.26 in Chapter 4. We can break out the energy equation in two-dimensional Cartesian coordinates. For starters, the left-hand side is nothing but the material derivative of temperature multiplied by density and specific heat at constant pressure:

$$\rho c_p \left(\frac{\partial T}{\partial t} + \vec{V} \cdot \vec{\nabla} T \right) = \rho c_p \left(\frac{\partial T}{\partial t} + u\frac{\partial T}{\partial x} + v\frac{\partial T}{\partial y} + \underbrace{w\frac{\partial T}{\partial z}}_{\text{ignore in 2D}} \right) \tag{5.70}$$

$$= \rho c_p \left(\frac{\partial T}{\partial t} + u\frac{\partial T}{\partial x} + v\frac{\partial T}{\partial y} \right).$$

The conduction term is the Laplacian of temperature. We have seen the Laplacian before when we covered the incompressible Navier–Stokes equations in Chapter 3. The Laplacian of temperature in 2D Cartesian coordinates is:

$$k\nabla^2 T = k \left(\frac{\partial^2}{\partial x^2} + \frac{\partial^2}{\partial y^2} + \underbrace{\frac{\partial^2}{\partial z^2}}_{\text{ignore}} \right) T \tag{5.71}$$

$$= k \left(\frac{\partial^2 T}{\partial x^2} + \frac{\partial^2 T}{\partial y^2} \right).$$

The last term we will break out into 2D Cartesian coordinates is the viscous dissipation term. From Chapter 3, recall that the velocity gradient is given as:

$$\vec{\nabla}\vec{V} = \begin{pmatrix} \frac{\partial u}{\partial x} & \frac{\partial v}{\partial x} & \frac{\partial w}{\partial x} \\[2mm] \frac{\partial u}{\partial y} & \frac{\partial v}{\partial y} & \frac{\partial w}{\partial y} \\[2mm] \frac{\partial u}{\partial z} & \frac{\partial v}{\partial z} & \frac{\partial w}{\partial z} \end{pmatrix}$$

In two dimensions, the velocity gradient plus its transpose becomes (ignoring the z-component):

$$\vec{\nabla}\vec{V} + \left(\vec{\nabla}\vec{V}\right)^{\dagger} = \begin{pmatrix} \frac{\partial u}{\partial x} & \frac{\partial v}{\partial x} \\ \frac{\partial u}{\partial y} & \frac{\partial v}{\partial y} \end{pmatrix} + \begin{pmatrix} \frac{\partial u}{\partial x} & \frac{\partial u}{\partial y} \\ \frac{\partial v}{\partial x} & \frac{\partial v}{\partial y} \end{pmatrix} = \begin{pmatrix} 2\frac{\partial u}{\partial x} & \frac{\partial v}{\partial x} + \frac{\partial u}{\partial y} \\ \frac{\partial u}{\partial y} + \frac{\partial v}{\partial x} & 2\frac{\partial v}{\partial y} \end{pmatrix}. \tag{5.72}$$

We can calculate the double dot product of the equation above with the velocity gradient to get:

$$\mu\left(\vec{\nabla}\vec{V} + \left(\vec{\nabla}\vec{V}\right)^{\dagger}\right) : \vec{\nabla}\vec{V} = \mu \begin{pmatrix} 2\frac{\partial u}{\partial x} & \frac{\partial v}{\partial x} + \frac{\partial u}{\partial y} \\ \frac{\partial u}{\partial y} + \frac{\partial v}{\partial x} & 2\frac{\partial v}{\partial y} \end{pmatrix} : \begin{pmatrix} \frac{\partial u}{\partial x} & \frac{\partial v}{\partial x} \\ \frac{\partial u}{\partial y} & \frac{\partial v}{\partial y} \end{pmatrix}$$

$$= \mu \left[2\left(\frac{\partial u}{\partial x}\right)^2 + \underbrace{\left(\frac{\partial v}{\partial x} + \frac{\partial u}{\partial y}\right)\frac{\partial v}{\partial x} + \left(\frac{\partial u}{\partial y} + \frac{\partial v}{\partial x}\right)\frac{\partial u}{\partial y}}_{\text{this can be simplified}} + 2\left(\frac{\partial v}{\partial y}\right)^2 \right]. \tag{5.73}$$

Observe that the middle terms in Equation 5.73 simplify quite a bit:

$$\left(\frac{\partial v}{\partial x} + \frac{\partial u}{\partial y}\right)\frac{\partial v}{\partial x} + \left(\frac{\partial u}{\partial y} + \frac{\partial v}{\partial x}\right)\frac{\partial u}{\partial y} = \left(\frac{\partial v}{\partial x}\right)^2 + \left(\frac{\partial u}{\partial y}\right)\left(\frac{\partial v}{\partial x}\right) + \left(\frac{\partial u}{\partial y}\right)^2 + \left(\frac{\partial v}{\partial x}\right)\left(\frac{\partial u}{\partial y}\right)$$

$$= \left(\frac{\partial v}{\partial x}\right)^2 + 2\left(\frac{\partial u}{\partial y}\right)\left(\frac{\partial v}{\partial x}\right) + \left(\frac{\partial u}{\partial y}\right)^2$$

$$= \left(\frac{\partial u}{\partial y} + \frac{\partial v}{\partial x}\right)^2. \tag{5.74}$$

Introducing the simplified version of the middle terms given in Equation 5.74 into Equation 5.73 gives the following expression for the viscous dissipation term in 2D Cartesian coordinates:

$$\mu\left(\vec{\nabla}\vec{V} + \left(\vec{\nabla}\vec{V}\right)^{\dagger}\right) : \vec{\nabla}\vec{V} = \mu \left[2\left(\frac{\partial u}{\partial x}\right)^2 + \left(\frac{\partial u}{\partial y} + \frac{\partial v}{\partial x}\right)^2 + 2\left(\frac{\partial v}{\partial y}\right)^2 \right]. \tag{5.75}$$

Using Equations 5.70, 5.71, and 5.75 in Equation 5.69 leads to:

$$\rho c_p\left(\frac{\partial T}{\partial t} + \vec{V}\cdot\vec{\nabla}T\right) = k\nabla^2 T + \mu\left(\vec{\nabla}\vec{V} + \left(\vec{\nabla}\vec{V}\right)^{\dagger}\right):\vec{\nabla}\vec{V} + \dot{q}_{gen}$$

$$\rightarrow \rho c_p\left(\frac{\partial T}{\partial t} + u\frac{\partial T}{\partial x} + v\frac{\partial T}{\partial y}\right) = k\left(\frac{\partial^2 T}{\partial x^2} + \frac{\partial^2 T}{\partial y^2}\right)$$
$$+ \mu\left[2\left(\frac{\partial u}{\partial x}\right)^2 + \left(\frac{\partial u}{\partial y} + \frac{\partial v}{\partial x}\right)^2 + 2\left(\frac{\partial v}{\partial y}\right)^2\right] + \dot{q}_{gen}.$$

We can write this a little more clearly by denoting the viscous dissipation term with the symbol Φ (which is common in fluid mechanics textbooks) to give us the **energy equation for an incompressible flow with constant thermal conductivity in 2D**:

$$\rho c_p\left(\frac{\partial T}{\partial t} + u\frac{\partial T}{\partial x} + v\frac{\partial T}{\partial y}\right) = k\left(\frac{\partial^2 T}{\partial x^2} + \frac{\partial^2 T}{\partial y^2}\right) + \Phi + \dot{q}_{gen}$$

$$\text{where } \Phi = \mu\left(2\left(\frac{\partial u}{\partial x}\right)^2 + \left(\frac{\partial u}{\partial y} + \frac{\partial v}{\partial x}\right)^2 + 2\left(\frac{\partial v}{\partial y}\right)^2\right) \tag{5.76}$$

Again, we can label Equation 5.76:

Incompressible Energy Equation in 2D Cartesian coordinates (non-conservation form)

$$\rho c_p\left(\underbrace{\frac{\partial T}{\partial t}}_{\substack{\text{local time}\\\text{derivative}}} + \underbrace{u\frac{\partial T}{\partial x} + v\frac{\partial T}{\partial y}}_{\text{advective transport}}\right) = k\underbrace{\left(\frac{\partial^2 T}{\partial x^2} + \frac{\partial^2 T}{\partial y^2}\right)}_{\substack{\text{diffusive transport}\\\text{(conduction)}}} + \underbrace{\Phi}_{\substack{\text{viscous dissipation}\\\text{(frictional heating)}}} + \dot{q}_{gen}$$

$$\text{where } \Phi = \mu\left(2\left(\frac{\partial u}{\partial x}\right)^2 + \left(\frac{\partial u}{\partial y} + \frac{\partial v}{\partial x}\right)^2 + 2\left(\frac{\partial v}{\partial y}\right)^2\right).$$

Equation 5.76 is probably one of the more common forms of the energy equation you will see, especially in an engineering context. Notice that the viscous dissipation term (Φ) is always positive, and thus will contribute to increasing the temperature of the fluid.

5.4 Full Governing Equations of Fluid Motion

At this point we should take stock and write down the full governing equations of fluid motion (noting that any of the variations discussed for the energy equation are acceptable), written here in Lagrangian form for compressible flows:

Continuity equation:

$$\frac{D\rho}{Dt} + \rho\vec{\nabla} \cdot \vec{V} = 0$$

Navier–Stokes equations:

$$\rho\frac{D\vec{V}}{Dt} = -\vec{\nabla}p + \vec{\nabla}\left(-\frac{2}{3}\mu\vec{\nabla} \cdot \vec{V}\right) + \vec{\nabla} \cdot \left(\mu\vec{\nabla}\vec{V} + \mu\left(\vec{\nabla}\vec{V}\right)^{\dagger}\right) + \rho\vec{g}$$

Energy equation:

$$\rho c_p \frac{DT}{Dt} = \vec{\nabla} \cdot \left(k\vec{\nabla}T\right) - \frac{2}{3}\mu\left(\vec{\nabla} \cdot \vec{V}\right)^2$$
$$+ \mu\left(\vec{\nabla}\vec{V} + \left(\vec{\nabla}\vec{V}\right)^{\dagger}\right) : \vec{\nabla}\vec{V} + \beta_T T\frac{Dp}{Dt} + \dot{q}_{gen}$$

As we have discussed in a previous chapter, there are six unknowns for the compressible flow equations: three components of velocity, pressure, density, and temperature. We currently only have five equations: continuity, three equations for the individual components of the Navier–Stokes equations, and the energy equation. As mentioned before, we need an equation of state for the compressible flow equations in order to ensure that the number of equations equals the number of unknowns. We have stated that a possible equation of state could be the ideal gas equation of state:

Equation of state for an ideal gas

$$p = \rho RT.$$

There are other equations of states, of course, depending on the material but the ideal gas equation is used quite a bit to model gases such as air.

5.4.1 Incompressible Version of the Governing Equations

In situations where an incompressible flow can be assumed (with a constant dynamic viscosity and thermal conductivity), the equations become (in non-conservation form):

Incompressible continuity equation:

$$\vec{\nabla} \cdot \vec{V} = 0$$

Incompressible Navier–Stokes equations:

$$\rho\left(\frac{\partial \vec{V}}{\partial t} + \vec{V} \cdot \vec{\nabla}\vec{V}\right) = -\vec{\nabla}p + \mu\nabla^2\vec{V} + \rho\vec{g}$$

Incompressible energy equation with constant k:

$$\rho c_p\left(\frac{\partial T}{\partial t} + \vec{V} \cdot \vec{\nabla}T\right) = k\nabla^2 T + \mu\left(\vec{\nabla}\vec{V} + \left(\vec{\nabla}\vec{V}\right)^\dagger\right) : \vec{\nabla}\vec{V} + \dot{q}_{gen}$$

The incompressible version of the equations is very popular. As mentioned in Chapter 3, the energy equation is not needed to solve for the dynamics of the flow. For an incompressible flow, only the Navier–Stokes equations and the constraint that the velocity must be divergence free is needed to solve for the velocity field. For an incompressible flow, the velocity field is used as an input to the energy equation. The incompressible version of the Navier–Stokes equations is often cited as being the Navier–Stokes equations (as opposed to the compressible version). In part because it is very useful in engineering and physics applications as well as it is somewhat easier (albeit still very difficult) to understand. In addition, mathematicians have taken a liking to studying some of the properties of the incompressible version. In fact, proving the existence of smooth solutions that are globally defined for the three-dimensional incompressible Navier–Stokes equations is one of the seven Millennium Prize problems offered as a challenge by the Clay Mathematics Institute. The Institute has offered to pay one million US dollars to anyone who can solve that problem or find a counter-example.

5.4.2 Incompressible Governing Equations in 2D Cartesian Coordinates

The incompressible version of the governing equations in 2D Cartesian coordinates will be useful to memorize since they come up quite often when solving problems (the cylindrical version of the equations is also often used for calculations). The incompressible version of the governing equations in two dimensions are:

Incompressible continuity equation:

$$\frac{\partial u}{\partial x} + \frac{\partial v}{\partial y} = 0$$

x-momentum equation (with no gravity):

$$\rho\left(\frac{\partial u}{\partial t} + u\frac{\partial u}{\partial x} + v\frac{\partial u}{\partial y}\right) = -\frac{\partial p}{\partial x} + \mu\left(\frac{\partial^2 u}{\partial x^2} + \frac{\partial^2 u}{\partial y^2}\right)$$

y-momentum equation (with gravity):

$$\rho\left(\frac{\partial v}{\partial t} + u\frac{\partial v}{\partial x} + v\frac{\partial v}{\partial y}\right) = -\frac{\partial p}{\partial y} + \mu\left(\frac{\partial^2 v}{\partial x^2} + \frac{\partial^2 v}{\partial y^2}\right) + \rho g_y$$

Incompressible energy equation with constant k:

$$\rho c_p\left(\frac{\partial T}{\partial t} + u\frac{\partial T}{\partial x} + v\frac{\partial T}{\partial y}\right) = k\left(\frac{\partial^2 T}{\partial x^2} + \frac{\partial^2 T}{\partial y^2}\right) + \Phi + \dot{q}_{gen}$$

$$\text{where } \Phi = \mu\left(2\left(\frac{\partial u}{\partial x}\right)^2 + \left(\frac{\partial u}{\partial y} + \frac{\partial v}{\partial x}\right)^2 + 2\left(\frac{\partial v}{\partial y}\right)^2\right)$$

A commonality between the Navier–Stokes equations and the energy equation is that both have an advective type term as well as a diffusive term. We discussed the advective term in Chapter 2 but we have not said anything about diffusion. This will be the topic for Section 5.5.

5.5 Diffusion

If you have ever smelled a good home-cooked meal, chances are that diffusion played a significant role in getting the food molecules from the cooking pot to your nose. If the air is still and you don't "advect" (i.e., move) the food particles by waving your hand, they will diffuse and spread themselves out among the air molecules. Notice that, in general, the molecules that make up the food will eventually spread themselves out such that it will generally be equally likely to find them in any corner of the room. In other words, it is highly unlikely that all of the food molecules will find themselves all in one corner of the room at a given instant in time. This "spreading" out effect is one way to think about diffusion. The second law of thermodynamics is essentially a statement of this effect. As mentioned previously, a very common statement of the second law is that heat cannot transfer, by itself, from a cold body to a hot body. Thus, if you put two bodies in contact, one hot and the other cold, the cold body will not get colder while the hot body gets warmer. Instead, the hot body will cool down and the cold body will warm up until some common temperature is reached between the both of them (i.e., when the two are in thermal equilibrium). Thus, the temperature "information" of the two bodies "spreads" out, much like food molecules would spread out in a room.[5] The origin of this spreading effect stems from molecular interactions. In the case

---

[5] It should be mentioned that even though the second law of thermodynamics is historically (at least in classical thermodynamics) written in terms of heat and temperature, diffusion of any kind (including the diffusion of food particles) is considered to be a second law phenomenon.

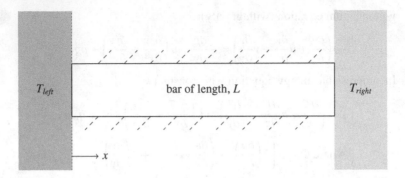

Figure 5.5 A bar of length, L, with the left end held at a temperature of T_{left} and the right end held at a temperature T_{right}.

of thermal diffusion (i.e., conduction), molecules that vigorously vibrate (and hence have a higher temperature) transfer their kinetic energy to molecules that have a more sluggish vibrational kinetic energy (and hence have a lower temperature). This transfer of vibrational kinetic energy of the molecules from a high to low vibrational kinetic energy is considered the thermal diffusion process. The same thing can be said for momentum diffusion. Molecules that have a higher momentum (in a given direction) bump into and transfer their momentum to molecules that have a smaller momentum. Momentum diffusion is modeled by the diffusive transport term of the Navier–Stokes equations.

To study the diffusion process, we are going to look at two different partial differential equations: Laplace's equation and the heat equation.

5.5.1 Steady-State Heat Conduction and the Laplace's Equation

We will further study diffusion with a simple example.

Example Consider the situation shown in Figure 5.5. This figure illustrates a bar of length L whose temperature at the left end (at $x = 0$) is held fixed at T_{left} and the temperature at the right end (at $x = L$) is held fixed at T_{right}.[6] Assume the energy transfer only occurs in the x-direction and all of the other sides (front, back, top, bottom) are insulated (i.e., no energy loss via heat occurs). At steady state, find the temperature distribution (i.e., $T(x)$) of the bar.

Solution Many of you will be able to deduce that the temperature of the bar as a function of x (i.e., the temperature distribution) will just be a straight line going from T_{left} at $x = 0$ to T_{right} at $x = L$. However, it might behoove us to just

---

[6] The mechanism for how the temperature at each end is held fixed is not important for the current discussion.

go through the mathematical machinery. Given that this is just a stationary bar, we can simplify our incompressible energy, Equation 5.69, by setting $\vec{V} = 0$. We can also set the time derivative to zero because we are in a steady state. In addition, there no heat generation, so $\dot{q}_{gen} = 0$. These assumptions give us (with thermal conductivity divided out):

$$\nabla^2 T = 0. \tag{5.77}$$

Equation 5.77 is a very well studied equation in physics, engineering, and mathematics. It is called Laplace's equation, named after the famous French scientist and mathematician Pierre-Simon Laplace. Laplace's equation will be the starting point for understanding the nature of diffusion.

By breaking Laplace's equation out in Cartesian coordinates and assuming that the only direction of importance is the x-direction, we get:

$$\nabla^2 T = \frac{\partial^2 T}{\partial x^2} + \cancel{\frac{\partial^2 T}{\partial y^2}} + \cancel{\frac{\partial^2 T}{\partial z^2}}^{\;0} = 0$$

$$\rightarrow \frac{d^2 T}{dx^2} = 0. \tag{5.78}$$

Laplace's equation, which is a partial differential equation, now becomes an ordinary differential equation in one dimension, hence the reason we switched the partial derivative (with the '∂') to an ordinary derivative with the 'd.'

Equation 5.78 is the governing equation for this particular problem. The general solution to this very simple differential equation, which can be obtained by integrating both sides of the equation twice with respect to x, is:

$$T = C_1 x + C_2. \tag{5.79}$$

The boundary conditions for the left ($x = 0$) side and the right ($x = L$) side can be written as such:

$$\text{at } x = 0, \quad T = T_{left}$$
$$\text{at } x = L, \quad T = T_{right}.$$

Applying the first boundary condition (i.e., at $x = 0$, $T = T_{left}$) makes our general solution, that is, Equation 5.79, become:

$$T_{left} = C_1 0 + C_2$$
$$\therefore C_2 = T_{left}.$$

Applying the right boundary condition, at $x = L, T = T_{right}$, leads to (with C_2 now set to T_{left}):

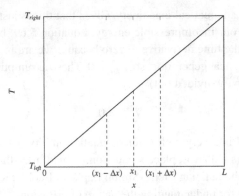

Figure 5.6　Plot of Equation 5.80 for the temperature in a bar.

$$T_{right} = C_1 L + T_{left}$$
$$\therefore C_1 = \frac{T_{right} - T_{left}}{L}.$$

Thus plugging in the expressions for C_1 and C_2 into Equation 5.79 yields a particular solution for this problem:

$$T = \frac{T_{right} - T_{left}}{L} x + T_{left} \tag{5.80}$$

This is the temperature distribution of the bar experiencing thermal diffusion (conduction) at steady state. Figure 5.6 provides an illustration of this temperature distribution. As you could have imagined, it is a straight line.

The temperature at any point along the bar is just the average of the temperature of neighboring equidistant points. To show this, consider two points a distance Δx away from x_1. One of the points is a distance Δx to the left of x_1 and the other is a distance Δx to the right of x_1, as shown in Figure 5.6. We can take the average temperature of the two points to get:

$$\frac{T\big|_{x=x_1+\Delta x} + T\big|_{x=x_1-\Delta x}}{2} =$$

$$= \frac{\left(\left(\frac{T_{right}-T_{left}}{L}\right)(x_1 + \Delta x) + T_{left}\right) + \left(\left(\frac{T_{right}-T_{left}}{L}\right)(x_1 - \Delta x) + T_{left}\right)}{2}$$

$$= \underbrace{\left(\frac{T_{right} - T_{left}}{L}\right) x_1 + T_{left}}_{\substack{\text{same as Equation 5.80} \\ \text{for } T \text{ of the bar at } x = x_1.}}$$

which, as you can see, is the same as the temperature at x_1. This is as expected due to the linearity of our solution.

Likewise, we could have considered a two-dimensional scenario with the following two-dimensional Laplace's equation as the governing equation:

$$\frac{\partial^2 T}{\partial x^2} + \frac{\partial^2 T}{\partial y^2} = 0.$$

In the two-dimensional case, the temperature of any point (x, y) is just the average of all the points that surround it some equal distance away. In other words, it would be the average temperature of a ring centered around point (x, y). Similarly, in the three-dimensional situation, the average temperature of the surface of a sphere is equal to the temperature at the center of the sphere.

This "averaging" behavior we see from the solutions to Laplace's equation provides an insight into the nature of diffusion. In simple terms, since each point is just the average value of its equidistant neighbors, you can think of Laplace's equation as having a solution that essentially "spreads" out information as much as possible so that there are no "hot spots" in the solution. Thus, the solution will appear smooth with no discontinuities.

In the above scenario, we assumed the system was in steady state, meaning that the temperature of the bar did not change with time. The next step to understanding diffusion can be accomplished by adding the local time derivative to Laplace's equation. The addition of the local time derivative will lead to another very famous partial differential equation: the heat equation. By studying the heat equation, the smoothing nature of diffusion will become even more apparent. In addition, the heat equation will illustrate the tendency of a diffusive system to approach an equilibrium state.

5.5.2 Unsteady Conduction and the Heat Equation

To study the effects of time on a diffusive system, consider the bar system similar to the one we studied in the last example, except this time the ends of the bar are held fixed at the same temperature, T_b. In addition, we are going to assume that the initial value of the bar (i.e., at $t = 0$) will be at temperature, T_i. This scenario is given in Figure 5.7.

We are going to assume that the bar is just a solid piece of material and, thus, has no associated fluid velocity ($\vec{V} = 0$). Therefore, we only care about the energy equation. In addition, we are going to ignore any heat generation ($\dot{q}_{gen} = 0$). After simplifying, the incompressible energy equation, Equation 5.69, becomes:

$$\rho c_p \frac{\partial T}{\partial t} = k\nabla^2 T \tag{5.81}$$

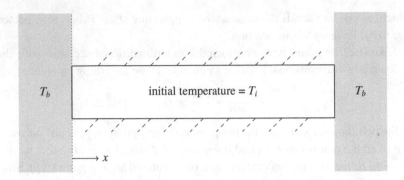

Figure 5.7 The setup of a problem used to study the heat equation.

We can divide out by ρc_p to get:

$$\frac{\partial T}{\partial t} = \frac{k}{\rho c_p}\nabla^2 T. \tag{5.82}$$

We are going to define a new, but common, material property called the thermal diffusivity, given by the symbol α:

$$\alpha = \frac{k}{\rho c_p}. \tag{5.83}$$

The thermal diffusivity can now be introduced into Equation 5.82 to give us:

$$\frac{\partial T}{\partial t} = \alpha\nabla^2 T. \tag{5.84}$$

Equation 5.84 is a partial differential equation known as the heat equation. The heat equation, sometimes referred to as the diffusion equation,[7] is another example of a well-studied partial differential equation in mathematics. Since the heat equation governs diffusion processes over time, the solutions of the heat equation typically "smooth" out as time passes.

If the unknown variable in the heat equation is temperature (as it is in this case), then the equation governs thermal diffusion (conduction). If the unknown variable is velocity, then the equation governs momentum diffusion. The unknown could also be concentration of a solute, in which case the heat equation would govern the diffusion of a solute in solution. An example of this might be the diffusion (or spreading) of a pinch of salt in water.

---

[7] You may see a few texts that make a distinction between the heat equation and the diffusion equation, with the diffusion equation being considered a touch more general. However, I use them interchangeably and I believe most authors do the same.

As mentioned, we have introduced a new material property, the thermal diffusivity (α). The thermal diffusivity determines how well a diffusion process occurs. The higher the α, the faster the diffusion process occurs. The units of α are meters squared per second, or $\frac{m^2}{s}$. To give you an idea of some numbers, the value of thermal diffusivity, under standard conditions, for water is $1.45 \times 10^{-7} \frac{m^2}{s}$, for air is $1.9 \times 10^{-5} \frac{m^2}{s}$, and for copper is $1.1 \times 10^{-4} \frac{m^2}{s}$.

We now turn to solving the heat equation for our specific scenario. We are going to assume that the only spatial direction of importance is the x-direction. Thus, the heat equation becomes the 1D heat equation:

$$\frac{\partial T}{\partial t} = \alpha \left(\frac{\partial^2 T}{\partial x^2} + \overset{0}{\cancel{\frac{\partial^2 T}{\partial y^2}}} + \cancel{\frac{\partial^2 T}{\partial z^2}} \right).$$

$$\rightarrow \frac{\partial T}{\partial t} = \alpha \frac{\partial^2 T}{\partial x^2}.$$

(5.85)

This situation requires two boundary conditions and one initial condition for time. The boundary and initial conditions can be written as:

$$\text{at } t = 0, \quad T = T_i \quad \text{(initial condition)}$$

$$\text{at } x = 0, \quad T = T_b \quad \text{(boundary condition)}$$

$$\text{at } x = L, \quad T = T_b \quad \text{(boundary condition)}$$

The solution to this partial differential equation problem requires the use of a method known as separation of variables. Even though separation of variables is a very standard technique for solving certain types of partial differential equations, instead of covering it we are just going to provide the solution. The solution to this problem is the following:

$$T = T_b + (T_i - T_b)\Sigma_{n=0}^{\infty} e^{-n^2\pi^2 Fo} \sin\left(\frac{n\pi}{L}x\right)$$

where Fo is a non-dimensional number known as the Fourier number and is given by:

$$Fo = \frac{\alpha t}{L^2}.$$

Note that T is a function of x and time (t), thus $T = T(x, t)$. The plot of the solution, when T_i is assumed to be larger than T_b, is given at various times in Figure 5.8. Notice that at $t = 0$ the temperature distribution is what you would expect: a horizontal line where the temperature is equal to the initial temperature. As time passes, the bar begins to cool down from the sides since T_b is assumed to be lower than T_i. The bar continually cools down until a steady state is reached where the temperature of the bar is uniformly T_b. The reason for this is because the $\frac{\partial^2 T}{\partial x^2}$ term will be negative all along the temperature

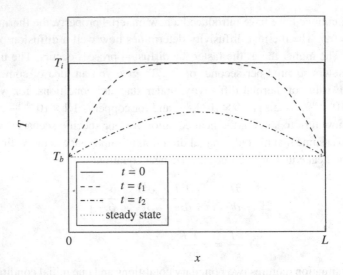

Figure 5.8 The plot of the temperature distribution at various times within the bar. Note that t_2 is greater than t_1 and steady state occurs at a time greater than t_2.

distribution for this scenario and hence will cause $\dfrac{\partial T}{\partial t}$ to be negative, thus dropping the temperature as time goes on. If T_b was hotter than T_i, then the opposite would occur and $\dfrac{\partial^2 T}{\partial x^2}$ would be positive, thus causing the temperature of the bar to increase until an equilibrium is obtained. This is the essence of the heat equation and diffusion in general. Diffusion seeks uniformity, or an equilibrium situation, whether it be thermal diffusion or momentum diffusion (as in the Navier–Stokes equations).

5.6 Convection–Diffusion Equation: Combined Advection and Diffusion

We now want to consider what happens when we combine the heat equation (which modeled diffusion) and the advection equation (which modeled advection). If we assume a situation of an incompressible flow with negligible viscous dissipation (we will illustrate when you can ignore viscous dissipation in the next chapter), then the 2D Cartesian coordinate version of the incompressible energy equation, that is, Equation 5.76, becomes:

$$\rho c_p \left(\frac{\partial T}{\partial t} + u \frac{\partial T}{\partial x} + v \frac{\partial T}{\partial y} \right) = k \left(\frac{\partial^2 T}{\partial x^2} + \frac{\partial^2 T}{\partial y^2} \right) + \dot{q}_{gen}$$

We can divide out by ρc_p from the left-hand side to get:

$$\frac{\partial T}{\partial t} + u\frac{\partial T}{\partial x} + v\frac{\partial T}{\partial y} = \alpha\left(\frac{\partial^2 T}{\partial x^2} + \frac{\partial^2 T}{\partial y^2}\right) + \frac{\dot{q}_{gen}}{\rho c_p} \tag{5.86}$$

where again recall that $\alpha = \dfrac{k}{\rho c_p}$. This is a variation of an equation known as the convection–diffusion equation, sometimes also called the advection-diffusion equation. This version is for 2D Cartesian coordinates but the name also applies to the general case using the del operator as:

$$\frac{\partial T}{\partial t} + \vec{V}\cdot\vec{\nabla}T = \alpha\nabla^2 T + \frac{\dot{q}_{gen}}{\rho c_p} \tag{5.87}$$

In order to illustrate what this equation does, it will be easier for us to use an even simpler version. We are going to make the additional assumption of one-dimension (i.e., $v = 0$) and zero heat generation. This makes Equation 5.86:

$$\frac{\partial T}{\partial t} + u\frac{\partial T}{\partial x} = \alpha\frac{\partial^2 T}{\partial x^2} \tag{5.88}$$

We have already studied various pieces of the convection–diffusion equation. For example, if we set the right-hand side of Equation 5.88 to zero, we are left with the advection equation for temperature as studied in Chapter 2:

$$\frac{\partial T}{\partial t} + u\frac{\partial T}{\partial x} = 0 \tag{2.16}$$

If we set the second term on the left-hand side of Equation 5.88 to zero, we are left with the heat equation (for 1D) we just discussed in the previous section, that is, Equation 5.85.

We know from Chapter 2 that the advection equation "moves" (or advects) the initial temperature profile with velocity u. It does not distort the shape of the temperature profile. On the other hand, the heat equation does not advect the temperature profile but instead "smooths" out the temperature profile via diffusion and pushes the system towards equilibrium.

The result of solving the convection–diffusion equation would behave as you might expect: the temperature profile advects and diffuses at the same time. Take, for example, a situation where the initial temperature distribution is a top-hat, such as that given in Figure 5.9. Applying Equation 5.88 to this initial temperature distribution with a velocity of u to the right would cause the temperature distribution to move to the right and smooth out at the same time. It would continue to do so until it eventually flattens completely at some point downstream (assuming the right boundary is far enough downstream to allow for this to happen). Figure 5.9 provides an illustration of this process.

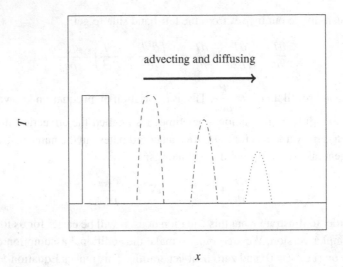

Figure 5.9 A sketch illustrating the application of the convection–diffusion equation in one dimension, that is, Equation 5.88, to an initial top hat temperature distribution. The velocity direction of the flow is assumed to be to the right.

The diffusion term $\left(\text{i.e., the } \alpha\dfrac{\partial^2 T}{\partial x^2} \text{ term}\right)$ causes the distribution to tend to an equilibrium (i.e. it causes the temperature to reach a final uniform value). The advective term $\left(\text{i.e., the } u\dfrac{\partial T}{\partial x} \text{ term}\right)$ contributes to moving the curve downstream with a velocity u. The larger the velocity, the more significant the advective term, and the further downstream the curve travels before completely flattening out. On the other hand, the larger the thermal diffusivity, the more significant the diffusion term and the quicker the curve will flatten.

The convection–diffusion equation models a process known as **convection**. It should be pointed out that the definition of convection may vary depending on the field of study. For example, in mechanical engineering, convection is usually considered to be a combination of advection and diffusion. Thus advection is modeled using the advection equation and convection is modeled using the convection–diffusion equation. In meteorology, convection is often distinguished from advection in that convection is usually used to denote the vertical transport of a property (such as energy) in air that is moving due to temperature differences and advection is the horizontal transport of a property. In other fields, convection and advection might be used interchangeably. Whatever the definition used, the main idea is the same: convection, much like advection, involves the transport of a quantity due to fluid motion.

5.6.1 The Viscous Burgers' Equation

The convection–diffusion equation shares some similarities with the Navier–Stokes equations, in particular, the incompressible Navier–Stokes equations.

If we look at the incompressible Navier–Stokes equations, Equation 3.53, with the pressure term ignored, we get:

$$\rho\left(\frac{\partial \vec{V}}{\partial t} + \vec{V} \cdot \vec{\nabla}\vec{V}\right) = \underbrace{-\vec{\nabla}p}_{\text{ignore}} + \mu\nabla^2\vec{V} + \rho\vec{g}$$

$$\rightarrow \rho\left(\frac{\partial \vec{V}}{\partial t} + \vec{V} \cdot \vec{\nabla}\vec{V}\right) = \mu\nabla^2\vec{V} + \rho\vec{g}$$

Dividing out by ρ leads to:

$$\frac{\partial \vec{V}}{\partial t} + \vec{V} \cdot \vec{\nabla}\vec{V} = \nu\nabla^2\vec{V} + \vec{g} \tag{5.89}$$

where ν is a momentum diffusivity, also known as the kinematic viscosity and has units of $\frac{m^2}{s}$. It is related to the dynamic viscosity (μ) via:

$$\nu = \frac{\mu}{\rho}.$$

If we are only interested in one dimension, such as the x-direction, then Equation 5.89 becomes (ignoring gravity):

$$\frac{\partial u}{\partial t} + u\frac{\partial u}{\partial x} = \nu\frac{\partial^2 u}{\partial x^2} \tag{5.90}$$

Equation 5.90 is called the viscous Burgers' equation. We studied the inviscid Burgers' equation in Chapter 2. The difference now is that the viscous version has a diffusion term $\left(\text{i.e., the } \nu\frac{\partial^2 u}{\partial x^2} \text{ term}\right)$. This diffusion term can help mitigate the development of shock waves that occurred when we studied the inviscid Burgers' equation.

This competition and tug-of-war between the diffusion term and the advective term occur in the Navier–Stokes equations as well as the energy equation. We will explore this interplay between diffusion and advection in the next section.

5.7 The Boundary Layer

We can consider a situation that will be very illustrative of the interplay between the diffusion and advection terms. Consider a fluid with a velocity U_∞ and a temperature T_∞ moving through a heated section (such as a hot screen), causing the fluid to reach a temperature of T_1 at the screen location. To set

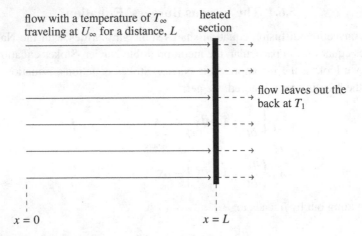

Figure 5.10 The setup to investigate the interplay between advection and diffusion in a purely thermal situation.

some boundary markers, suppose the "start" of the flow begins at $x = 0$ and the heated section is located at $x = L$. This situation is depicted in Figure 5.10. We are going to assume an incompressible flow situation with a completely uniform velocity in the x-direction of U_∞. We are going to make the additional assumptions of steady state and that the temperature will only vary in the x-direction. With these assumptions, we can cross off the following terms in the 2D Cartesian coordinate energy equation, that is, Equation 5.76:

$$\rho c_p \left(\cancel{\frac{\partial T}{\partial t}} + \underbrace{U_\infty}_{=u} \frac{\partial T}{\partial x} + \cancel{v \frac{\partial T}{\partial y}} \right) = k \left(\frac{\partial^2 T}{\partial x^2} + \cancel{\frac{\partial^2 T}{\partial y^2}} \right) + \mu \cancel{\left(\vec{\nabla}\vec{V} + \left(\vec{\nabla}\vec{V}\right)^\dagger \right) : \vec{\nabla}\vec{V}}.$$

The time derivative goes away because of the steady state assumption, the viscous dissipation goes away because of the uniform velocity assumption, the advective term in the y-direction is eliminated because the y-velocity is zero, and the y-component of the diffusive term goes away because of the assumption that the temperature only varies with the x-direction. Note that we also will assume no heat generation. Thus, we are left with the following equation:

$$\rho c_p U_\infty \frac{dT}{\partial x} = k \frac{d^2 T}{dx^2} \tag{5.91}$$

Dividing out by $\rho c_p U_\infty$ leads to (with again $\alpha = \frac{k}{\rho c_p}$):

$$\frac{dT}{\partial x} = \frac{\alpha}{U_\infty} \frac{d^2 T}{dx^2} \tag{5.92}$$

We have the following two boundary conditions:

$$\text{at } x = 0, \quad T = T_\infty$$
$$\text{at } x = L, \quad T = T_1.$$

Solving Equation 5.92 with the above boundary conditions leads to:[8]

$$T = T_\infty + (T_1 - T_\infty)\frac{\exp\left(Pe\frac{x}{L}\right) - 1}{\exp(Pe) - 1}. \tag{5.93}$$

The exponents in Equation 5.93 appear to have a parameter we have not seen before, that is, the Pe. The Pe is a non-dimensional (i.e., it has no units) parameter called the Péclet number. The Péclet number is given by:

$$Pe = \frac{U_\infty L}{\alpha}. \tag{5.94}$$

Non-dimensional parameters show up quite often in fluid mechanics. They are usually a ratio of different physical phenomena at play in a problem. For example, the Péclet number can be thought of as a ratio of advection to conduction effects. To see this, consider writing the advection term and diffusion term as algebraic equations, that is:

$$\text{Advection term: } \rho c_p U_\infty \frac{dT}{dx} \to \rho c_p U_\infty \frac{\Delta T}{L}$$

$$\text{Diffusion term: } k\frac{d^2 T}{dx^2} = k\frac{d}{dx}\left(\frac{dT}{dx}\right) \to k\frac{1}{L}\left(\frac{\Delta T}{L}\right). \tag{5.95}$$

where ΔT is a change in some temperature (for example, in this problem, $\Delta T = T_1 - T_\infty$) and L is the change in x. Notice that the second order derivative in the diffusive transport term is broken out into a derivative of a derivative. This notation makes it clear that the temperature change is not squared in the numerator (as I sometimes see students itching to do).

By dividing out the "algebraic" advective transport term with the "algebraic" conduction term we get the Péclet number, observe:

$$Pe = \frac{\text{advection}}{\text{conduction}} = \frac{\rho c_p U_\infty \frac{\Delta T}{L}}{k\frac{1}{L}\left(\frac{\Delta T}{L}\right)} = \frac{\rho c_p U_\infty L}{k} = \frac{U_\infty L}{\alpha}. \tag{5.96}$$

The thought process we used to get to the result of Equation 5.96 will be useful for Chapter 6 when we discuss nondimensionalization. For the moment, we can take a look at the result of the problem at different Péclet numbers to see if we can glean any useful information about the physics. To do so, consider the

---

[8] We will not bother to go through the machinery of solving this equation because it is not necessary for our discussion.

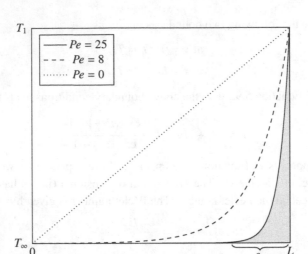

Figure 5.11 Temperature distribution from Equation 5.93 plotted at various Péclet numbers. The shaded region is the thermal boundary layer for $Pe = 25$. A boundary layer is a region where diffusion begins to play a role in an otherwise advection-dominated situation. The approximate boundary layer thickness, δ_T, for the $Pe = 25$ case is also shown. The boundary layer for the $Pe = 8$ case is larger than the $Pe = 25$ case. The $Pe = 0$ situation is not considered to have a boundary layer because it is a pure diffusion (i.e., conduction) case.

plot of Equation 5.93 given in Figure 5.11 for various Péclet numbers. Notice what happens when Pe is zero, the solution becomes a straight-line temperature distribution from T_∞ to T_1.[9] This result looks exactly like the result we got when discussing pure conduction through a bar. Getting a result similar to a pure conduction scenario makes sense considering that the Péclet number is zero, which means the ratio of advection to conduction is zero, which implies a situation that is dominated by conduction. Another approach to thinking about it is the following: since the Péclet is zero, you can visualize this as meaning the velocity of the flow (i.e., U_∞) is zero. A zero flow velocity means that only conduction is involved in the problem, since nothing is advecting.

Next, consider the situation when the Péclet number becomes large. Examining Figure 5.11, we can assume that the temperature distribution would stay "flatter" (i.e., closer to the value of the incoming flow temperature, T_∞) for a greater distance until it absolutely must increase to reach T_1 in order to satisfy the boundary condition at $x = L$. What is this telling us? One way to visualize

---

[9] Note, in order to see this, you will have to perform l'Hôpital's rule since setting Pe to zero leaves Equation 5.93 indeterminant with the result being $\frac{0}{0}$.

what is happening here is to assume a large incoming flow velocity (which would imply a large Péclet number as long as L is not too small and α is not too large). A large flow velocity would mean that the temperature of the incoming flow, since it is moving fast, would not have time to change until it absolutely had to in order to meet the boundary temperature requirements on the right end at $x = L$. This means it would stay closer to the value of T_∞ for a longer distance. Thus, a large Péclet number (which is considered an advection dominated problem rather than a conduction one) would mean the temperature of the flow would be mostly advected (i.e., moved) without any change in value until a point downstream where the fluid begins to "sense" that it needs to increase in temperature in order to meet the value of T_1. The moment when the fluid begins to "sense" when to start increasing is the point where conduction effects start to become important in a flow that is otherwise just happily advecting temperature. This point is the start of what is called the boundary layer. A **boundary layer** is a region of a flow where diffusion becomes important in a situation that is otherwise dominated by advection. Figure 5.11 shades the boundary layer region for a situation when $Pe = 25$. The boundary layer thickness is dependent on the value of the Péclet number. The larger the Péclet number, the smaller the boundary layer thickness. As you can tell, the boundary layer thickness for the situation when $Pe = 8$ would be larger than the $Pe = 25$ case. In addition, the $Pe = 0$ is not considered to even have a boundary layer since it is pure diffusion (conduction). As the Péclet number approaches infinity, the boundary layer thickness would approach zero, in which case the temperature value would need to jump to T_1 at $x = L$, forming a discontinuity.

We can actually use the algebraic ratio of advection to conduction to estimate the thickness of the boundary layer. To do so, consider that the start of the boundary layer is the region in which diffusion becomes important in an advection-dominated situation (i.e., when advection and diffusion start to roughly become of equal importance). If this is the case, then we can set the algebraic form for the advection term to the algebraic form for the diffusion term via:

$$\underbrace{\rho c_p U_\infty \frac{\Delta T}{\delta_T}}_{\text{advection}} = \underbrace{k \frac{1}{\delta_T} \left(\frac{\Delta T}{\delta_T} \right)}_{\text{conduction}}$$

where δ_T is the boundary layer thickness (note the subscript "T" in δ_T indicates we are dealing with a temperature, or thermal, boundary layer). Notice we are using the boundary layer thickness instead of L for the length scale because the region where diffusion and advection begin to compete has a size of δ_T, not L. After all, the situation is mostly just advecting for a large distance until it reaches the boundary layer. Most of the "action" (i.e., temperature change) takes place at the start of the boundary layer. We can solve for δ_T:

$$\delta_T = \frac{k}{\rho c_p} \frac{1}{U_\infty} = \frac{\alpha}{U_\infty}. \tag{5.97}$$

Usually, thermal boundary layer thicknesses are given in terms of an overall Péclet number. Therefore, by multiplying the right hand side of Equation 5.97 by $\frac{L}{L}$ (where L is the overall length of the problem), we get:

$$\delta_T = \frac{\alpha}{U_\infty} \frac{L}{L} = \frac{\alpha}{U_\infty L} L$$

$$= \frac{L}{Pe}$$

As you can see, the higher the Pe number, the smaller the thermal boundary layer thickness. Conversely, the lower the Pe number, the larger the boundary layer thickness. It should be noted that this analysis just provided an approximate relationship for the size of the boundary layer.

Reynolds Number

Another nondimensional number that comes up all of the time (and is the most important of all nondimensional numbers in fluid mechanics) is the Reynolds number. Much like the Péclet number, which provides information on the overall balance between advection and diffusion for the energy equation, the Reynolds number provides the overall relationship between advection and diffusion in the Navier–Stokes equations. In the Navier–Stokes equations, an "algebraic" version of the advection and diffusion terms can be given as (only the x-direction advection and diffusion terms of the incompressible u velocity is used here):

$$\text{Advection term: } \rho u \frac{\partial u}{\partial x} \to \rho U_\infty \frac{U_\infty}{L}$$

$$\text{Diffusion term: } \mu \frac{\partial^2 u}{\partial x^2} = \mu \frac{\partial}{\partial x} \left(\frac{\partial u}{\partial x} \right) \to \mu \frac{1}{L} \left(\frac{U_\infty}{L} \right).$$

This assumes the overall velocity change is U_∞ and the change in x is L. Taking the ratio of advection to diffusion gives us the Reynolds number:

$$Re = \frac{\text{advection}}{\text{diffusion}} = \frac{\rho U_\infty \frac{U_\infty}{L}}{\mu \frac{1}{L} \left(\frac{U_\infty}{L} \right)} = \frac{\rho U_\infty L}{\mu}.$$

The Reynolds number can also be written using kinematic viscosity (i.e., momentum diffusivity, $\nu = \frac{\mu}{\rho}$):

$$Re = \frac{\rho U_\infty L}{\mu} = \frac{U_\infty L}{\nu}.$$

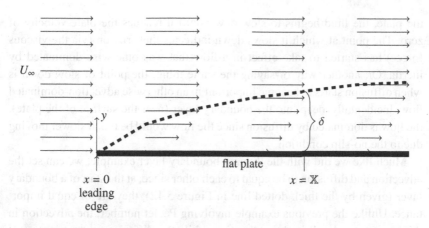

Figure 5.12 The setup for flow past a flat plate. This is the prototypical problem for studying boundary layers. The boundary layer starts at the thick dashed line. The arrows denote the velocity vectors (there actually should be a slight tilt upwards in the velocity vector inside the boundary layer since there is a small y-velocity component). The boundary layer thickness increases in size further downstream from the leading edge of the plate.

Note also that in many fluid mechanics textbooks, the Reynolds number is considered a ratio of inertia to viscous forces.

5.7.1 Boundary Layers and the Reynolds Number

We discussed the idea of a boundary layer in the last section. We used an example that involved the energy equation to solve for a temperature distribution for a given scenario that involved a nondimensional parameter called a Péclet number. Such a boundary layer, since it involved temperature, is often called a thermal (or temperature) boundary layer and is used quite often by those who study thermal convection. Another example of a boundary layer can be illustrated with the very standard fluid mechanics problem of flow past a flat plate. This problem is illustrated in Figure 5.12. This is, in fact, the prototypical example illustrating a boundary layer. In this problem, the flow moves past the flat plate with a velocity of U_∞. As it passes the leading edge, the flow near the plate slows down to match the velocity of the plate (which is typically zero since the plate is normally stationary) due to the no-slip condition discussed in the last chapter. A velocity profile (i.e., a y vs. u) plot can be made at various points along the plate.

The velocity far above the plate, which is called the free stream velocity, stays a constant U_∞. Thus, far above the plate, there is only advection and is governed by the inviscid Burgers' equation. However, as you move closer to the plate, the fluid begins to slow down until it reaches the plate velocity of zero. The point at which it slows down indicates that friction (via the viscous forces) has started to take effect in a flow that was otherwise dominated by inertia. Or, another way of saying the same thing, the point of slow down is when diffusion starts to become important in an otherwise advection-dominated flow. Incidentally, deep into the boundary layer (near the surface of the plate), the flow is dominated by diffusion since the flow would be much slower moving due to the no-slip condition.

Much like we did with the thermal boundary layer example, we can set the advection and diffusion to be equal to each other since, at the start of a boundary layer (given by the thick dotted line in Figure 5.12), they are of equal importance. Unlike the previous example involving Péclet number, the advection in this case is nearly all in the x-direction while the diffusion is most important in the y-direction since it is diffusion that is causing the velocity to change with respect to the y-direction. The length scales associated with the x- and y-directions are the distance downstream from the leading edge (X) and the boundary layer thickness (δ), respectively. Hence, the "algebraic" advection (which is dominant in the x-direction) would look something like this:

$$\text{advection} = \rho \underbrace{\frac{U_\infty^2}{X}}_{\rho u \frac{\partial u}{\partial x}}$$

while the "algebraic" diffusion (which is dominant in the y-direction) is:

$$\text{diffusion} = \mu \underbrace{\frac{1}{\delta} \frac{U_\infty}{\delta}}_{\mu \frac{\partial^2 u}{\partial y^2}}$$

Setting the diffusion equal to the advection (which is done because the boundary layer starts at the point where the two effects are of equal importance) would give an expression for δ:

$$\rho \frac{U_\infty^2}{X} = \mu \frac{1}{\delta} \frac{U_\infty}{\delta}$$

$$\rightarrow \delta = \sqrt{\frac{\mu X}{\rho U_\infty}}.$$

If we define the Reynolds number for this case to be:

$$Re_x = \frac{\rho U_\infty X}{\mu}.$$

Then the boundary layer thickness can be written as $\left(\text{after multiplying by } \frac{X}{X}\right)$:

$$\delta = \sqrt{\frac{\mu X}{\rho U_\infty} \frac{X}{X}} = \sqrt{\frac{\mu}{\rho U_\infty X}} X = \frac{X}{\sqrt{Re_x}}.$$

It is important to note that this is an estimate for the boundary layer thickness. Notice that this expression implies the boundary layer thickness grows as we move further downstream. In addition, notice that the boundary layer shrinks as the Reynolds number increases. These things will be explored some more in the next chapter, where the concept of nondimensionalization (sometimes referred to as scaling) will be introduced. In addition, note we have not mentioned anything about turbulent flow (i.e., chaotic flow). The analysis we have been doing only considers laminar (or regular and predictable) flow.

5.8 Boundary Conditions for the Energy Equation

At this point, we would like to begin thinking about how we would solve the energy equation for some simple problems. Before we do so, however, it would be good to revisit boundary conditions. The boundary conditions for the energy equation come in similar flavors as the boundary conditions for the Navier–Stokes equations. They are:

(i) **Dirichlet boundary condition**: This type of boundary condition occurs when a specified temperature is given at a particular boundary. For example:

$$\text{at } y = 0.1 \text{ m, } T = 1000 \text{ K}$$

We have already used Dirichlet boundary conditions when we covered the heat equation and Laplace's equation.

(ii) **Neumann boundary condition**: This boundary condition, often called a heat flux boundary condition in the context of the energy equation, specifies the value of the derivative of temperature at a specified boundary. This boundary condition is used often in the energy equation because it is common to specify a heat flux at a given boundary. If you recall, the heat flux is given by: $\vec{q}'' = -k\vec{\nabla}T$, which contain derivatives of temperature through the $\vec{\nabla}$ operation. An example of this type of boundary condition for the energy equation might look like the following:

$$\text{at } x = 0.5 \text{ m, } \vec{q}'' \cdot \vec{n} = 100 \ \frac{\text{W}}{\text{m}^2}$$

where $\vec{n}$ is the outward normal of the boundary surface. If the surface is pointed in the positive x-direction, then $\vec{n} = \hat{\imath}$ and the boundary condition above becomes:

$$\text{at } x = 0.5 \text{ m}, \quad -k\left(\hat{i}\frac{\partial T}{\partial x} + \hat{j}\frac{\partial T}{\partial y}\hat{k}\frac{\partial T}{\partial z}\right) \cdot \hat{i} = 100 \; \frac{\text{W}}{\text{m}^2}$$

$$\rightarrow \text{at } x = 0.5 \text{ m}, \quad -k\frac{\partial T}{\partial x} = 100 \; \frac{\text{W}}{\text{m}^2}.$$

(iii) **Robin boundary condition**: In the context of the energy equation, this type of boundary condition is sometimes called a convection boundary condition due to the fact that it comes up in the study of thermal convection (which, incidentally, is heavily reliant on fluid mechanics). It utilizes both the value of the dependent variable in question as well as its derivatives. An example of this boundary condition might look like this:

$$\vec{q}'' \cdot \vec{n}\bigg|_{x=0} = h\left(T\bigg|_{x=0} - T_\infty\right)$$

where h is a constant (usually called the heat transfer coefficient or the convection coefficient) and T_∞ is a constant value typically detailing a known temperature of the fluid at some far distant point.

In the next section, we are going to revisit the two examples we did in the last chapter (i.e., Couette flow and pressure-driven flow) with the inclusion of the energy equation.

5.9 Examples

5.9.1 Temperature Profile in a Couette Flow

We are going to now look at the energy equation in a Couette flow, or shear-driven flow, with a plate moving to the right at U_T, as shown in Figure 5.13. It is a similar situation as discussed in Chapter 4 except now there is some temperature information provided. The temperature of the bottom plate is T_B and the temperature of the top plate is T_T.

We are interested in the **temperature profile (i.e., the temperature as a function of y) of the fluid between the plates at steady state**.

We will start with the incompressible energy equation and make some assumptions. The assumptions we will make will be: steady state ($\frac{\partial T}{\partial t} = 0$), no heat generation ($\dot{q}_g = 0$), velocity in the y- and z-directions are negligible ($v = w = 0$), the derivatives in the z-direction is ignored, and the velocity is fully developed ($\frac{\partial u}{\partial x} = 0$). The fully developed assumption was discussed in the last chapter when we solved for the velocity profile. This leads to a simplification of Equation 5.76 via:

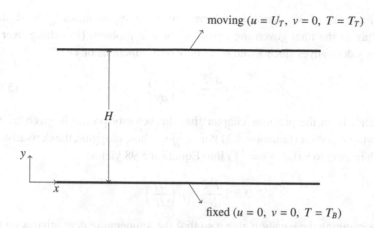

Figure 5.13 Schematic of Couette flow with temperature included.

$$\rho c_p \left(\frac{\partial T}{\partial t} + u \frac{\partial T}{\partial x} + v \frac{\partial T}{\partial y} \right) = k \left(\frac{\partial^2 T}{\partial x^2} + \frac{\partial^2 T}{\partial y^2} \right) + \dot{q}_g + \Phi$$

$$\text{where } \Phi = \mu \left[2 \left(\frac{\partial u}{\partial x} \right)^2 + 2 \left(\frac{\partial v}{\partial y} \right)^2 + \left(\frac{\partial u}{\partial y} + \frac{\partial v}{\partial x} \right)^2 \right].$$

After the simplifications, we have:

$$\rho c_p u \frac{\partial T}{\partial x} = k \left(\frac{\partial^2 T}{\partial x^2} + \frac{\partial^2 T}{\partial y^2} \right) + \mu \left(\frac{\partial u}{\partial y} \right)^2.$$

This equation can be simplified further by noting that $\frac{\partial T}{\partial x} = 0$. The main reason for this is that the wall temperatures (T_B and T_T) are constants. This implies that the temperatures at the wall do not vary with x. Since the temperatures of the walls are constant, and hence do not vary with x, and also noting that the velocity does not change with x, we are going to assume that the temperature of the flow, internally, will also not change with x. Thus, $\frac{\partial T}{\partial x} = 0$. Is this a good assumption? In some sense, maybe not, given that the motion of the fluid would increase the internal temperature through the viscous dissipation (frictional heating) term, which would then get advected downstream (i.e., transported to the right) by the fluid motion. Thus, potentially increasing the temperature of the fluid downstream. However, the fact that the temperatures of the walls are constant ensures, at least for slow-moving flows, that diffusion between the walls and the fluid will cause any slight increase in temperature of the flow downstream due to advecting the heat caused by viscous dissipation to quickly adjust so that minimal temperature change occurs downstream.

As a result of our discussion above, we are going to make $\frac{\partial T}{\partial x} = 0$, thus resulting in the final governing equation for this problem (switching over to ordinary derivatives since T and u are now only functions of y):

$$0 = k\frac{d^2T}{dy^2} + \mu\left(\frac{du}{dy}\right)^2. \tag{5.98}$$

Recall from the previous chapter that the velocity profile is given by the following equation (Equation 4.31): $u = \frac{U_T}{H}y$. Thus, plugging the derivative of u with respect to y (i.e., $\frac{du}{dy} = \frac{U_T}{H}$) into Equation 5.98 yields:

$$0 = k\frac{d^2T}{dy^2} + \mu\left(\frac{U_T}{H}\right)^2.$$

Rearranging the equation above so that the temperature derivative is on the left leads to (after dividing out k):

$$\frac{d^2T}{dy^2} = -\frac{\mu}{k}\left(\frac{U_T}{H}\right)^2.$$

The general solution to the above differential equation can be obtained by integrating both sides with respect to y twice to get:

$$T = -\frac{\mu}{2k}\left(\frac{U_T}{H}\right)^2 y^2 + C_3 y + C_4. \tag{5.99}$$

The boundary conditions to be applied are:

$$\text{at } y = 0, \quad T = T_B$$
$$\text{at } y = H, \quad T = T_T.$$

Applying the first boundary condition (e.g., at $y = 0$, $T = T_B$) to our general solution leads to:

$$T_B = -\frac{\mu}{2k}\left(\frac{U_T}{H}\right)^2 0^2 + C_3 0 + C_4$$
$$\therefore C_4 = T_B.$$

Applying the next boundary condition (e.g., at $y = H$, $T = T_T$) gives:

$$T_T = -\frac{\mu}{2k}\left(\frac{U_T}{H}\right)^2 H^2 + C_3 H + \underbrace{C_4}_{T_B}$$

$$\therefore C_3 = \frac{1}{H}\left(T_T - T_B + \frac{\mu}{2k}\left(\frac{U_T}{H}\right)^2 H^2\right).$$

With the constants of integrations now known, our equation for temperature is:

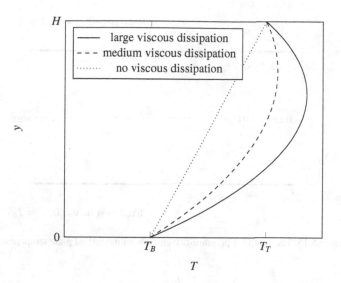

Figure 5.14 The temperature profile for Couette flow with constant plate temperatures.

$$T = -\frac{\mu}{2k}\left(\frac{U_T}{H}\right)^2 y^2 + \frac{1}{H}\left(T_T - T_B + \frac{\mu}{2k}\left(\frac{U_T}{H}\right)^2 H^2\right)y + T_B. \quad (5.100)$$

The temperature profile (i.e., the y vs. T curve) is plotted in Figure 5.14. Notice that when the velocity of the top is zero, the temperature profile becomes a straight line. This should make sense because a zero plate velocity implies that there is no flow at all and thus there would be no viscous dissipation, which would make the temperature profile purely dependent on conduction (diffusion). However, if the velocity of the top is non-zero, then the temperature of the flow is higher (relative to the pure diffusion situation of $U_T = 0$), as illustrated in Figure 5.14. The reason behind this is that the viscous dissipation term contributes only in a positive way (due to the fact that the dissipation term is squared). Thus, it will contribute positively to the temperature. Again, this is much like frictional heating where rubbing your hands together will only increase the temperature of your hands and never decrease the temperature of your hands. So, the larger the viscous dissipation term (as generated by the moving fluid), the higher the temperature the fluid would be.

5.9.2 Temperature Profile in Pressure-driven Channel Flow

We are now going to consider temperature in pressure-driven flow between two parallel plates. We have already obtained the velocity profile from Chapter 4, as can be seen in Equation 4.34 as shown below for reference:

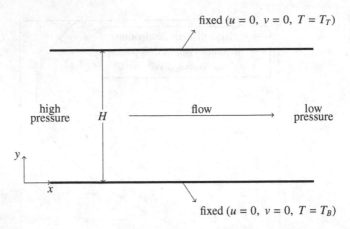

Figure 5.15 The setup for pressure-driven flow with constant plate temperatures.

$$u = \frac{1}{2\mu} \frac{dp}{dx} \left(y^2 - Hy \right).$$

Now, we are given the temperature of the top plate is T_T and the temperature of the bottom plate is T_B, as shown in Figure 5.15.

The energy equation for this problem simplifies to the exact same simplified form in our lid-driven Couette flow, that is, Equation 5.98:

$$0 = k \frac{d^2 T}{dy^2} + \mu \left(\frac{du}{dy} \right)^2 \tag{5.98}$$

This time, however, the velocity derivative in our dissipative term is no longer a constant since the derivative of the velocity profile with respect to y, Equation 4.34, becomes an equation for a line:

$$\frac{du}{dy} = \frac{1}{2\mu} \frac{dp}{dx} (2y - H).$$

Plugging the derivative above into Equation 5.98 yields:

$$0 = k \frac{d^2 T}{dy^2} + \mu \left(\frac{1}{2\mu} \frac{dp}{dx} (2y - H) \right)^2.$$

Let's rearrange this equation and foil out the squared term to get:

$$\frac{d^2 T}{dy^2} = -\frac{1}{4k\mu} \left(\frac{dp}{dx} \right)^2 \left(4y^2 - 4Hy + H^2 \right).$$

Integrating both sides twice with respect to y yields the following general solution for T:

$$T = -\frac{1}{4k\mu}\left(\frac{dp}{dx}\right)^2\left(\frac{1}{3}y^4 - \frac{2}{3}Hy^3 + \frac{1}{2}H^2y^2\right) + C_3y + C_4 \qquad (5.101)$$

where C_3 and C_4 are constants of integration to be obtained from applying the boundary conditions:

$$\text{at } y = 0, \ T = T_B$$
$$\text{at } y = H, \ T = T_T.$$

Applying the first boundary condition (at $y = 0$, $T = T_B$) is simple enough and leads to $C_4 = T_B$.

Applying the second boundary condition (at $y = H$, $T = T_T$) is a little messier algebraically but fairly straightforward as well:

$$T_T = -\frac{1}{4k\mu}\left(\frac{dp}{dx}\right)^2\left(\frac{1}{3}H^4 - \frac{2}{3}HH^3 + \frac{1}{2}H^2H^2\right) + C_3H + \underbrace{T_B}_{=C_4}$$

$$\therefore C_3 = \frac{T_T - T_B}{H} + \frac{1}{24k\mu}\left(\frac{dp}{dx}\right)^2 H^3.$$

Thus, the resulting equation for the temperature profile after plugging in the constants of integration is:

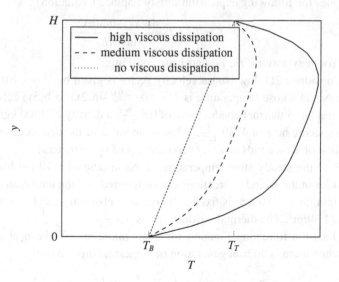

Figure 5.16 The temperature profile for a pressure-driven flow.

$$T = -\frac{1}{4k\mu}\left(\frac{dp}{dx}\right)^2\left(\frac{1}{3}y^4 - \frac{2}{3}Hy^3 + \frac{1}{2}H^2y^2\right)$$

$$+ \left(\frac{T_T - T_B}{H} + \frac{1}{24k\mu}\left(\frac{dp}{dx}\right)^2 H^3\right)y + T_B$$

(5.102)

Equation 5.102 is plotted in Figure 5.16. Like in the Couette flow case, the higher the viscous dissipation, the larger the temperature in between the plates.

In the next chapter, we are going to revisit many of the problems from the current chapter, except a new step will be added: nondimensionalization.

Problems

5.1 Evaluate the viscous dissipation term at an (x, y) location of $(1, 5)$ meters with a dynamic viscosity of 10^{-3} Pa·s if velocity field is given by: $\vec{V} = 2y\cos(2x)\hat{i} + y^2\sin(2x)\hat{j}$ m/s.

5.2 Calculate the power (per volume) if the stress tensor is:

$$\vec{\vec{T}} = \begin{pmatrix} 10x & 5y & 6x \\ 5y & 2y & 25y \\ 6x & 25y & 3z \end{pmatrix}$$

with a velocity vector is $\vec{V} = \hat{i} + 2\hat{j} - 2\hat{k}$ and no body force.

5.3 Does the following expression satisfy Laplace's equation?

$$T = T_0 + T_1 \frac{\sin(\pi x)\sinh(\pi y)}{\sinh(\pi L)}$$

You may assume T_0, T_1, and L are constants.

5.4 Consider a 2D flow whose velocity vector is given by $\vec{V} = -10x\hat{i} + 10y\hat{j}$ m/s and whose temperature is $T = 50e^{-0.03t}\sin(2x)\cosh(5y)$ kelvin. The fluid has a thermal conductivity of 0.6 $\frac{W}{mK}$, a density of 1000 kg/m³, and a specific heat of 4180 $\frac{J}{kgK}$. What is the value of the heat generation at a time of 10 seconds and (x, y) coordinates of $(1, 1)$ meters?

5.5 Find the steady state temperature in the middle of a 10 cm bar whose sides in the y- and z-directions can be ignored and the temperature of the right (at $x = 10$ cm) is fixed at 0 degrees Celsius and the flux on the left is 1 W/m². The thermal conductivity is 0.6 $\frac{W}{mK}$.

5.6 Does the following function for temperature satisfy the heat equation when there is no heat generation (α is thermal diffusivity):

$$T = T_0 + T_m \exp\left(\frac{-2\pi^2\alpha}{L^2}t\right)\sin\left(\frac{\pi x}{L}\right)\sin\left(\frac{\pi y}{L}\right)$$

You may assume L, T_0, and T_m are also constant.

5.7 Consider a system of water with a pinch of salt centered in the middle. If we were to model the "stirring" of the system of water using the advection equation, what would the result be? What would we need to do in order to more accurately model the physical reality of the situation?

5.8 Consider incompressible flow between two parallel plates. The flow is driven by the bottom plate moving to the right with a velocity of 100 m/s. The viscosity of the fluid is given by 0.001 Pa·s and the thermal conductivity is 0.6 $\frac{W}{mK}$. If both the bottom and the top temperatures are fixed at 0°C, what is the maximum temperature of the flow? You may assume the flow is in a steady-state and is only in the x-direction. In addition, you may ignore any pressure gradient as well as any heat generation. How is the maximum temperature of the flow different than the plate temperatures? What accounts for the difference?

5.9 In your own words, what is the definition of a boundary layer?

6

Nondimensionalization and Scaling

This chapter is somewhat of a deviation from the rest of the book and potentially could have been cut as it is not necessarily about the fluid equations per se. However, the concepts of this chapter, while the most complicated of the whole book, are arguably the most useful in understanding the various concepts of fluid mechanics. In addition, the concepts discussed within this chapter can be extended to other areas of physics, particularly areas that are overly reliant on differential equations (which is most of physics and engineering).

In this chapter, we are going to discuss a concept known as scaling. Scaling (also known as nondimensionalization) is essentially a form of dimensional analysis. **Dimensional analysis** is a general term used to describe a means of analyzing a system based on the units (e.g., kilogram for mass, kelvin for temperature, meter for length, coulomb for electric change, etc.) of the problem.

One aspect of scaling analysis is that although there is a general procedure, there is still a decent amount of decision-making during the analysis. Thus, scaling analysis can be considered somewhat of an art form since less "experienced" decisions might not lead to as fruitful of a result. The mastering of scaling analysis, however, can help the scientist or engineer better understand the problem at hand or greatly simplify the analysis.

6.1 The Idea Behind Nondimensionalization

We have actually already been introduced, albeit subtly, to the idea of nondimensionalization when we discussed the Reynolds number and the Peclet number in the last chapter. The Reynolds number and Péclet number are nondimensional parameters that provide a way of deciphering the relative importance of advection and diffusion in a given problem. The Reynolds number is

defined as a ratio of advection to diffusion for momentum, or similarly a ratio
of inertia to viscous forces, i.e.:

$$Re = \frac{\rho U_\infty L}{\mu} = \frac{U_\infty L}{\nu} \quad \left(\frac{\text{inertia}}{\text{viscous forces}} \quad \text{or} \quad \frac{\text{momentum advection}}{\text{momentum diffusion}} \right)$$

where U_∞ is considered to be a general velocity value used in the problem
(called a characteristic velocity scale) and L is the overall length value of the
problem (called a characteristic length scale). The Reynolds number is defined,
in general, using both definitions provided above.

Similarly, the Péclet number (which in many ways can be considered a thermal Reynolds number), is defined as the ratio of energy advection to thermal
diffusion (or conduction). i.e.:

$$Pe = \frac{U_\infty L}{\alpha} \quad \left(\frac{\text{energy advection}}{\text{thermal diffusion}} \quad \text{or} \quad \frac{\text{energy advection}}{\text{conduction}} \right)$$

Péclet and Reynolds number are just two examples of nondimensional parameters that provide key information on a problem. Other nondimensional
parameters show up in fluid mechanics as well. Such parameters are a way of
"collapsing" all of the physical variables in a problem (such as dynamic viscosity, density, the specific heat, the thermal conductivity, as well as the overall
velocity and length scale) into a much smaller set of parameters.

The general concept of reducing the parameter space into a smaller set
of nondimensional variables is one of the major uses of dimensional analysis. In this chapter, we will discuss a type of dimensional analysis called
nondimensionalization, or scaling. **Scaling** involves transforming differential
equations that contain dimensions (such as meters, seconds, kilograms, kelvin)
into nondimensional equations that contain a much smaller set of physical
parameters.

6.2 The Basics of Scaling Analysis

At this point, we will now provide the basic steps (to the extent there are steps)
for the scaling (nondimensionalization) procedure:

Step 1: Simplify the problem and equations (and boundary/initial conditions) as much as possible.

Step 2: Introduce scaled (i.e., nondimensional) variables and plug them into
the simplified equations and conditions.

Step 3: Divide the whole equation by a dimensional "coefficient." This dimensional coefficient is usually determined to be a coefficient in front of one
of the terms in the equation. If there are multiple terms in the equation and

each one has a coefficient in front, it is usually advised to divide by the biggest coefficient.

Step 4: Make any additional simplifications with the new scaled variables if necessary.

Step 5: Set any characteristic scale factors not previously defined such that the coefficient terms are of order one (this is the hardest part and will be explained as we go). Note, this may not show up in all problems.

Step 6: Solve the resulting equation if applicable.

Now we will use these steps with on some problems.

6.2.1 First Example: Conduction in a Wall with Known Wall Temperatures

In our first scaling example, we are going to re-investigate the problem of finding the steady state temperature distribution within a bar, where the left end (at $x = 0$) of the bar has a temperature fixed at T_{left} and the right end (at $x = L$) of the bar is held fixed at T_{right}. We know from the last chapter that the result for the temperature distribution is as shown in Equation 5.80, repeated here for reference:

$$T = \frac{T_{right} - T_{left}}{L}x + T_{left}. \tag{5.80}$$

This implies that the temperature within the bar is a function of the spatial variable x as well as three additional parameters: T_{left}, T_{right}, and L. Thus:

$$T = f(x, T_{left}, T_{right}, L).$$

So even for this very simple problem, the dependent variable we care about, which is temperature, has four total parameters: the independent coordinate variable (x) and three additional parameters (T_{left}, T_{right}, and L). Is it possible to establish a temperature expression with a reduced set of parameters? That is what we will try to do with scaling analysis.

To scale (i.e., nondimensionalize) this problem, we can follow the steps defined in the previous section (although not all steps will be applicable).

Step 1: Start with a general governing equation for the problem and simplify as much as possible (including the boundary/initial conditions).

To do that for this problem, since we are dealing with a pure one-dimensional conduction problem at steady state, our governing equation is nothing more than the conduction term in the x-direction equaling zero:

$$\frac{d^2T}{dx^2} = 0 \quad \text{or} \quad \frac{d}{dx}\left(\frac{dT}{dx}\right) = 0. \tag{6.1}$$

Notice we are breaking out the second derivative explicitly into a derivative of a derivative. The reason for doing this will become clear shortly in our second step.

Our boundary conditions are:

$$\text{at } x = 0, \quad T = T_{left}$$
$$\text{at } x = L, \quad T = T_{right}. \tag{6.2}$$

Step 2: Introduce scaled (or nondimensional) variables and plug them into the equations and conditions. This step will look a little odd at first but it is essentially nothing but ridding ourselves of the burden of dealing with arbitrary units such as meters, kelvin, etc.

For this problem, we have two variables we need to scale: our dependent variable that we want to solve for (temperature) and an independent variable (the spatial variable, x). Note, it is not necessary to scale the other parameters as they are constant values. To start, we can scale x. Note that the x variable has units of length (i.e., meters, feet, inches, etc.). In order to nondimensionalize x, we need to divide it by another parameter related to the problem at hand that also has a unit of length. It is probably clear that the only parameter in this problem that has a unit of length is L. So, we can divide x by L to get a scaled version of x, the scaled version of x will be denoted as such with a superscript $*$. Thus:

$$x^* = \frac{x}{L} \tag{6.3}$$

The L is the bar length and is also considered the characteristic length scale of the problem. Characteristic meaning it is the general length scale pertaining to the problem at hand. The variable, x^* is called the nondimensional x or the scaled x. It is a unitless variable.

Scaling temperature usually has a slightly different look than the scaling of x. In general, when scaling temperature, or any variable for that matter, one of the aspects, at least for some common problems, is to scale in such a way that the scaled problem becomes easier to handle. For this problem, we are going to scale the temperature in the following manner:

$$\theta^* = \frac{T - T_{left}}{T_{right} - T_{left}} \tag{6.4}$$

where θ^* is a nondimensional (or scaled) temperature.[1]

---

[1] We could have defined the nondimensional temperature to be the variable T^* but the standard variable used for the nondimensional temperature is θ^*.

You might be wondering what we are doing here with this temperature scaling. For starters, you may have noticed that we are subtracting T_{left} from T, which we did not do when we scaled x. We will call T_{left} a "shift factor" and, as we will see, utilizing a shift factor will simplify our problem. You also might be wondering why our "scale factor" (that is, the term in the denominator, i.e., $T_{left} - T_{right}$), is not just either T_{left} or T_{right}. In theory, we could have used either T_{left} or T_{right} as a scale factor. However, again as we will see, using the difference will allow for a simpler nondimensional problem.

So, we now have introduced our nondimensional variables, θ^* and x^*, let's plug them into our simplified governing equation, Equation 6.1. To do so, we need to convert the derivatives (and values from the boundary condition) in Equation 6.1 to derivatives involving the scaled variables.

So, let's take a look first at converting a first-order derivative of temperature to a derivative using scaled variables. An easy approach for doing this (which only works for constant shift and scale factors) is to plug our scaled variables directly into the derivative equation. In other words, solve Equations 6.3 and 6.4 for x and T, respectively, and plug into a derivative to get:

$$\frac{dT}{dx} = \frac{d\left(\overbrace{\left(T_{left} - T_{right}\right)\theta^* + T_{left}}^{T}\right)}{d\underbrace{\left(Lx^*\right)}_{x}}$$

T_{left} is a constant

$$= \frac{d\left(\left(T_{left} - T_{right}\right)\theta^*\right)}{d\left(Lx^*\right)} + \frac{d\left(T_{left}\right)}{d\left(Lx^*\right)} \quad \text{(distributed the derivative)}$$

$$= \underbrace{\frac{\left(T_{left} - T_{right}\right)}{L}}_{\substack{\text{pulled out of derivative} \\ \text{because they are constants}}} \frac{d\theta^*}{dx^*}.$$

Therefore, the derivative of T with respect to x can be written in terms of a derivative of the nondimensional temperature, θ^*, with respect to the nondimensional x, x^*, as:

$$\frac{dT}{dx} = \frac{\left(T_{left} - T_{right}\right)}{L} \frac{d\theta^*}{dx^*}. \tag{6.5}$$

We can do a similar procedure for the second derivative:

$$\frac{d^2T}{dx^2} = \frac{d}{dx}\left(\frac{dT}{dx}\right) = \frac{d}{d\left(Lx^*\right)}\left(\underbrace{\frac{\left(T_{left} - T_{right}\right)}{L}\frac{d\theta^*}{dx^*}}_{\frac{dT}{dx}}\right)$$

$$= \underbrace{\frac{\left(T_{left} - T_{right}\right)}{L^2}}_{\text{pulled out of derivative}}\underbrace{\frac{d}{dx^*}\left(\frac{d\theta^*}{dx^*}\right)}_{\frac{d^2\theta^*}{dx^{*2}}}.$$

Therefore, the second-order derivative becomes:

$$\frac{d^2T}{dx^2} = \frac{\left(T_{left} - T_{right}\right)}{L^2}\frac{d^2\theta^*}{dx^{*2}}. \tag{6.6}$$

Note, a more formal procedure to scale the derivatives is to use the chain-rule. However, what we did above is fine to do for constant-valued scale and shift factors (which is typical).

We can now introduce Equation 6.6 into Equation 6.1 to get:

$$\frac{\left(T_{left} - T_{right}\right)}{L^2}\frac{d^2\theta^*}{dx^{*2}} = 0. \tag{6.7}$$

Next we need to introduce Equation 6.3 and 6.4 into our boundary conditions, i.e., Equation 6.2, to get:

$$\text{at } x = 0, \quad T = T_{left} \quad \rightarrow \quad \text{at } \underbrace{Lx^*}_{x} = 0, \quad \underbrace{\left(T_{right} - T_{left}\right)\theta^* + T_{left}}_{T} = T_{left}$$

$$\text{at } x = L, \quad T = T_{right} \quad \rightarrow \quad \text{at } \underbrace{Lx^*}_{x} = L, \quad \underbrace{\left(T_{right} - T_{left}\right)\theta^* + T_{left}}_{T} = T_{right}.$$

$$\tag{6.8}$$

Step 3: Equation 6.7 contains the non-dimensional variables, θ^* and x^*, yet it is still not considered a nondimensional equation. The reason that it is not non-dimensional is because the units of the equation is degree of temperature per unit length squared, such as kelvin per meter squared ($\frac{K}{m^2}$). In order to make the equation nondimensional, or a scaled equation, we need to divide out by the $\frac{\left(T_{left} - T_{right}\right)}{L^2}$ factor, which will give us the scaled governing equation:

$$\frac{d^2\theta^*}{dx^{*2}} = 0. \tag{6.9}$$

Likewise, the boundary conditions are also not yet nondimensional. We can first consider the boundary condition at $x = 0$ (or $Lx^* = 0$):

$$\text{at } \underbrace{Lx^* = 0}_{\substack{\text{divide} \\ \text{by } L}}, \quad \underbrace{\left(T_{right} - T_{left}\right)\theta^* + T_{left} = T_{left}}_{\text{subtract by } T_{left} \text{ and divide by } \left(T_{right} - T_{left}\right)}$$

$$\rightarrow x^* = 0, \quad \theta^* = \underbrace{\frac{T_{left} - T_{left}}{T_{right} - T_{left}}}_{=0}.$$

Scaling the next boundary condition looks like this:

$$\text{at } \underbrace{Lx^* = L}_{\substack{\text{divide} \\ \text{by } L}}, \quad \underbrace{\left(T_{right} - T_{left}\right)\theta^* + T_{left} = T_{right}}_{\text{subtract by } T_{left} \text{ and divide by } \left(T_{right} - T_{left}\right)}$$

$$\rightarrow x^* = 1, \quad \theta^* = \underbrace{\frac{T_{right} - T_{left}}{T_{right} - T_{left}}}_{=1}.$$

Thus the scaled boundary conditions are:

$$\begin{aligned} \text{at } x^* = 0, \quad \theta^* = 0 \\ \text{at } x^* = 1, \quad \theta^* = 1. \end{aligned} \tag{6.10}$$

Notice that Equations 6.9 and 6.10 have been simplified and do not contain any parameters. This can be readily seen by comparing the geometry of the scaled (i.e., nondimensional) problem to the unscaled (i.e., dimensional) problem. This is done in Figure 6.1.

Steps 4 and 5 are not applicable for this particular example, so we will go to step 6.

Step 6: Solving Equation 6.9 for θ^* leads to the general solution of:

$$\theta^* = C_1 x^* + C_2. \tag{6.11}$$

Applying our scaled boundary conditions to our scaled general solution leads to:

$$\theta^* = x^* \tag{6.12}$$

which is just a simple line, as you would expect. Notice that our scaled temperature, θ^*, is only a function of x^*. This is a significant reduction in the parameter space compared to the unscaled (i.e., dimensional) temperature, $T = f(x, L, T_{left}, T_{right})$. To rescale, just reintroduce the scaling (from Equations 6.3 and 6.4) into Equation 6.12 to get:

$$\underbrace{\frac{T - T_{left}}{T_{right} - T_{left}}}_{\theta^*} = \underbrace{\frac{x}{L}}_{x^*}$$

$$\xrightarrow{\text{rearrange}} T = \frac{T_{right} - T_{left}}{L}x + T_{left}. \quad \text{(same as Equation 5.80)}$$

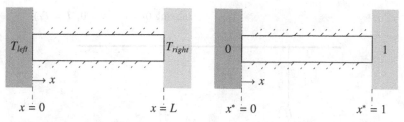

(a) Unscaled (dimensional) scenario.   (b) Scaled (nondimensional) scenario.

Figure 6.1 Comparison of the dimensional steady state conduction problem to the nondimensional (scaled) steady state conduction problem. Notice there are no longer the parameters T_{left}, T_{right}, and L in the scaled scenario.

We can continue practicing nondimensionalization with some additional familiar problems.

6.3 Couette Flow Revisited with Nondimensionalization

Let's take a look at Couette flow using scaling analysis. Again, as a reminder of Couette flow, consider the situation given in Figure 6.2. This is the same situation as we studied earlier with flow between two plates where the top plate is moving with a constant velocity of U_T and the bottom plate is fixed. Also, the temperature of the top plate is given by T_T and the bottom plate temperature is given by T_B. Can we find an expression for a nondimensional (i.e., scaled) velocity profile and temperature profile?

If we wanted, we can scale the result we obtained earlier when we solved for the velocity and temperature profile of Couette flow. However, we are instead going to redo this problem for the scaled velocity and temperature profiles.

Step 1: The simplified x-momentum and energy equation for this problem was obtained in Chapter 5:

$$\frac{d^2u}{dy^2} = 0 \qquad (x\text{-momentum})$$

$$\frac{d^2T}{dy^2} = -\frac{\mu}{k}\left(\frac{du}{dy}\right)^2 \qquad (\text{energy}).$$

There is no y-momentum equation for this problem because we are going to assume only a one-dimensional flow.

Our boundary conditions are:

$$\text{at } y = 0, \quad u = 0 \quad \text{and} \quad T = T_B$$
$$\text{at } y = H, \quad u = U_T \quad \text{and} \quad T = T_T.$$

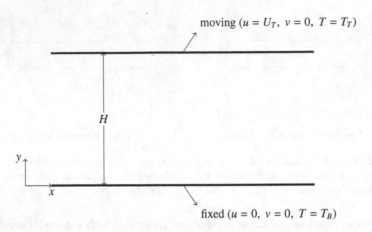

moving ($u = U_T$, $v = 0$, $T = T_T$)

H

y

x

fixed ($u = 0$, $v = 0$, $T = T_B$)

Figure 6.2 Schematic of Couette flow with temperature included.

Step 2: Introduce nondimensional variables:

$$y^* = \frac{y}{H}, \quad u* = \frac{u}{U_T}, \quad \theta^* = \frac{T - T_B}{T_T - T_B}.$$

Plugging these scaled variables into our simplified governing equations give:

$$\frac{U_T}{H^2} \frac{\partial^2 u^*}{\partial y^{*2}} = 0 \qquad \text{(x-momentum)}$$

$$\frac{T_T - T_B}{H^2} \frac{\partial^2 \theta^*}{\partial y^{*2}} = -\frac{\mu U_T^2}{k H^2} \left(\frac{\partial u^*}{\partial y^*} \right)^2 \quad \text{(energy).}$$

(6.13)

Next we will plug the scaled variables into the boundary conditions, i.e.:

$$\text{at } Hy^* = 0, \quad U_T u^* = 0 \quad \text{and} \quad (T_T - T_B)\theta^* + T_B = T_B$$
$$\text{at } Hy^* = H, \quad U_T u^* = U_T \quad \text{and} \quad (T_T - T_B)\theta^* + T_B = T_T.$$

(6.14)

Step 3: The equations in Equation 6.13 are not scaled yet as they still have units associated with them. To scale them, each one needs to be divided by a coefficient in front of one of the terms. In the case of the x-momentum equation, the only coefficient is the $\frac{U_T}{H^2}$. In the case of the temperature equation, there are two terms. The term on the left (the diffusion term) has a coefficient of $\frac{T_T - T_B}{H^2}$ in front of it and the term on the right (the viscous dissipation term) has the coefficient $\frac{\mu U_T^2}{k H^2}$. For now, we can divide the energy equation by either of these coefficients. We will pick the left coefficient and divide out by $\frac{T_T - T_B}{H^2}$. Thus, dividing out the x-momentum equation by $\frac{U_T}{H^2}$ and the energy by $\frac{T_T - T_B}{H^2}$ yields the following result for the scaled governing equations:

$$\frac{d^2 u^*}{dy^{*2}} = 0 \qquad \text{(scaled } x\text{-momentum)}$$

$$\frac{d^2 \theta^*}{dy^{*2}} = -\frac{\mu U_T^2}{k(T_T - T_B)} \left(\frac{du^*}{dy^*} \right)^2 \qquad \text{(scaled energy).}$$

(6.15)

The boundary conditions from Equation 6.14 also need to be scaled by dividing out by a coefficient. We can take a look at the $y = 0$ boundary conditions:

$$\text{at } \underbrace{Hy^* = 0}_{\text{divide by } H}, \quad \underbrace{U_T u^* = 0}_{\text{divide by } U_T} \quad \text{and} \quad \underbrace{(T_T - T_B)\theta^* + T_B = T_B}_{\text{subtract by } T_B \text{ and divide by } (T_T - T_B)}$$

$$\rightarrow \text{at } y^* = 0, \quad u^* = 0 \quad \text{and } \theta^* = \underbrace{\frac{T_B - T_B}{T_T - T_B}}_{=0}.$$

Now consider the $y = H$ boundary conditions:

$$\text{at } \underbrace{Hy^* = H}_{\text{divide by } H}, \quad \underbrace{U_T u^* = U_T}_{\text{divide by } U_T} \quad \text{and} \quad \underbrace{(T_T - T_B)\theta^* + T_B = T_T}_{\text{subtract by } T_B \text{ and divide by } (T_T - T_B)}$$

$$\rightarrow \text{at } y^* = 1, \quad u^* = 1 \quad \text{and } \theta^* = \underbrace{\frac{T_T - T_B}{T_T - T_B}}_{=1}.$$

Therefore, the scaled boundary conditions are:

$$\text{at } y^* = 0, \quad u^* = 0 \quad \text{and} \quad \theta^* = 0$$

$$\text{at } y^* = 1, \quad u^* = 1 \quad \text{and} \quad \theta^* = 1.$$

(6.16)

Step 4: We are now going to make some simplifications to the coefficient seen in the nondimensional temperature equation, i.e., the $\frac{\mu U_T^2}{k(T_T - T_B)}$ term in Equation 6.15. This coefficient can actually be written in terms of two named nondimensional numbers, the Eckert number (Ec) and the Prandtl number (Pr). The Eckert number is defined as:

$$Ec = \frac{U_T^2}{c_p (T_T - T_B)}.$$

The Eckert number is considered a ratio of the kinetic energy of the flow (i.e., ρU_T^2) to the enthalpy of the flow ($\rho c_p (T_T - T_B)$).

The Prandtl number is the ratio of kinematic viscosity (momentum diffusivity), v, to thermal diffusivity, α:

$$Pr = \frac{v}{\alpha}.$$

The Prandtl number is considered a material parameter since it is highly dependent on the type of material.

Recalling that the kinematic viscosity is related to the dynamic viscosity via $v = \mu/\rho$ and the thermal diffusivity is related to thermal conductivity via $\alpha = k/(\rho c_p)$, we can also write the Prandtl number as:

$$Pr = \frac{\mu/\rho}{k/(\rho c_p)} = \frac{\mu c_p}{k}.$$

Multiplying the Eckert number with the Prandtl number gives:

$$PrEc = \frac{\mu c_p}{k} \frac{U_T^2}{c_p (T_T - T_B)} = \frac{\mu U_T^2}{k (T_T - T_B)}.$$

which is exactly the coefficient in the energy equation in Equation 6.15. Thus, the energy equation can now be written as:

$$\frac{d^2\theta^*}{dy^{*2}} = -PrEc \left(\frac{du^*}{dy^*}\right)^2.$$

So, the final scaled equations and boundary conditions we are to solve are:

$$\frac{\partial^2 u^*}{\partial y^{*2}} = 0 \qquad \text{(scaled } x\text{-momentum)}$$

$$\frac{\partial^2 \theta^*}{\partial y^{*2}} = -PrEc \left(\frac{\partial u^*}{\partial y^*}\right)^2 \quad \text{(scaled energy).}$$

(6.17)

With boundary conditions:

$$\text{at } y^* = 0 : \quad u^* = 0 \quad \text{and} \quad \theta^* = 0$$
$$\text{at } y^* = 1 : \quad u^* = 1 \quad \text{and} \quad \theta^* = 1.$$

The scaled geometry is given in Figure 6.3.

Step 5: This step is not applicable to this problem but will be introduced in the next example on pressure-driven flow.

Step 6: We can now solve this problem. The x-momentum has a general solution that is linear. That is:

$$u^* = C_1 y^* + C_2.$$

where C_1 and C_2 are the unknowns. Applying the boundary conditions for u^* leads to:

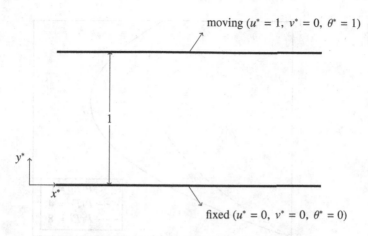

Figure 6.3 Schematic of the scaled Couette flow scenario.

$$\text{at } y^* = 0, \ u^* = 0 \ \rightarrow \ 0 = C_1 0 + C_2$$
$$\therefore C_2 = 0$$
$$\text{at } y^* = 1, \ u^* = 1 \ \rightarrow \ 1 = C_1 1 + 0$$
$$\therefore C_1 = 1.$$

So the final x-velocity profile is just a line given by:

$$u^* = y^*.$$

Next up is the energy equation. The energy equation requires the derivative of the x-velocity with respect to y, which is nothing but 1. Thus, the general solution to the energy equation is:

$$\frac{d^2\theta^*}{dy^{*2}} = -PrEc \underbrace{\left(\frac{du^*}{dy^*}\right)^2}_{=1} \xrightarrow{\text{solve}} \theta^* = -\frac{PrEc}{2}y^{*2} + C_3 y^* + C_4.$$

Applying the boundary conditions for the energy equation leads to:

$$\text{at } y^* = 0, \ \theta^* = 0 \ \rightarrow \ 0 = -\frac{PrEc}{2}0^2 + C_3 0 + C_4$$
$$\therefore C_4 = 0$$
$$\text{at } y^* = 1, \ \theta^* = 1 \ \rightarrow \ 1 = -\frac{PrEc}{2}1^2 + C_3 1$$
$$\therefore C_3 = 1 + \frac{PrEc}{2}.$$

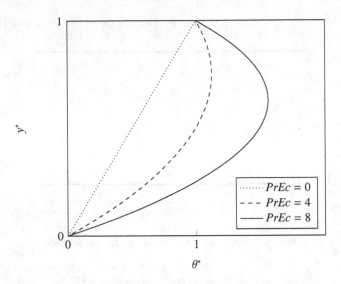

Figure 6.4 Scaled temperature profiles for Couette flow.

So, our final energy equation is (with constants C_3 and C_4 included):

$$\theta^* = -\frac{PrEc}{2}y^{*2} + \left(1 + \frac{PrEc}{2}\right)y^*.$$

Plots of the energy equation are given in Figure 6.4 for varying *PrEc* values. The larger the *PrEc* value, the more pronounced the viscous dissipation and therefore the higher the temperature is in between the parallel plates.

6.4 Pressure-driven Flow with Nondimensionalization

In this section we are going to revisit pressure-driven flow we studied at the end of Chapter 5. However, instead of a constant temperature at the bottom plate, we are going to assume the bottom plate is insulated (which results in the derivative of temperature in the *y*-direction at the bottom plate being zero). A schematic of this scenario is given in Figure 6.5. We want to find the scaled *x*-velocity profile as well as the scaled temperature profile.

Step 1: The simplified governing equations (obtained from earlier chapters) are:

$$\frac{d^2u}{dy^2} = \frac{1}{\mu}\frac{dp}{dx} \qquad (x\text{-momentum})$$

$$\frac{d^2T}{dy^2} = -\frac{\mu}{k}\left(\frac{du}{dy}\right)^2 \qquad (\text{energy}).$$

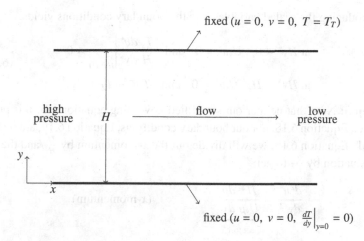

Figure 6.5 The setup for pressure-driven flow. Goal is to find the scaled *x*-velocity profile and the scaled temperature profile.

With boundary conditions:

$$\text{at } y = 0, \quad u = 0 \quad \text{and} \quad \frac{dT}{dy}\bigg|_{y=0} = 0$$

$$\text{at } y = H, \quad u = 0 \quad \text{and} \quad T = T_T.$$

Step 2: The scaling we are going to use for this problem will be:

$$y^* = \frac{y}{H}, \quad u* = \frac{u}{U_s}, \quad \theta^* = \frac{T - T_T}{T_s}$$

where H is the distance between the plates, U_s is considered to be the characteristic velocity scale in the *x*-direction whose expression will be determined later, and T_s is a characteristic temperature scale whose expression will also be determined later. You may be wondering why we are not using T_T as the characteristic temperature scale and the reason is simply that there is a slightly better option, as we will come to find out. Also notice that we did not scale the pressure gradient in this case because we are assuming the pressure gradient is just a constant value.

Plugging our scaling into the differential equations leads to:

$$\frac{U_s}{H^2} \frac{d^2 u^*}{dy^{*2}} = \frac{1}{\mu} \frac{dp}{dx} \qquad (x\text{-momentum})$$

$$\frac{T_s}{H^2} \frac{d^2 \Theta^*}{dy^{*2}} = -\frac{\mu U_s^2}{kH^2} \left(\frac{du^*}{dy^*}\right)^2 \qquad (\text{energy}).$$

(6.18)

Introducing the scaled variables into the boundary conditions yields:

$$\text{at } Hy^* = 0, \quad U_s u^* = 0 \quad \text{and} \quad \frac{T_s}{H}\frac{d\theta^*}{dy^*}\bigg|_{Hy^*=0} = 0 \tag{6.19}$$

$$\text{at } Hy^* = H, \quad U_s u^* = 0 \quad \text{and} \quad T_s\theta^* + T_T = T_T.$$

Step 3: Note that neither our simplified governing equations for this problem, i.e., Equation 6.18, nor our boundary conditions, Equation 6.19, are scaled. To scale Equation 6.18, we will divide out the x-momentum by $\frac{U_s}{H^2}$ and the energy equation by $\frac{T_s}{H^2}$ to get:

$$\frac{d^2u^*}{dy^{*2}} = \underbrace{\frac{H^2}{U_s\mu}\frac{dp}{dx}}_{\text{set to 1}} \qquad (x\text{-momentum})$$

$$\frac{d^2\theta^*}{dy^{*2}} = -\underbrace{\frac{\mu U_s^2}{kT_s}}_{\text{set to 1}}\left(\frac{du^*}{dy^*}\right)^2 \qquad (\text{energy}).$$

In addition, the boundary conditions can be rearranged to get:

$$\text{at } y^* = 0, \quad u^* = 0 \quad \text{and} \quad \frac{d\theta^*}{dy^*}\bigg|_{y^*=0} = 0 \tag{6.20}$$

$$\text{at } y^* = 1, \quad u^* = 0 \quad \text{and} \quad \theta^* = 0.$$

Steps 4 and 5: Notice that since the velocity scale (U_s) and temperature scale (T_s) have not been set. By setting the coefficients in the x-momentum and energy equations to 1 (which, at the moment, just makes our calculations nicer), we can find an expression for the U_s and T_s. The U_s becomes:

$$U_s = \frac{H^2}{\mu}\frac{dp}{dx} \tag{6.21}$$

and the T_s becomes:

$$T_s = \frac{\mu U_s^2}{k}. \tag{6.22}$$

You might be wondering why we included the pressure gradient in the coefficient. The reason is simply because the pressure gradient is just a constant (i.e., a number) and thus can be absorbed into the coefficient.

Notice that the T_s contains U_s in the expression. We can replace the U_s to get a final form for T_s:

$$T_s = \frac{\mu}{k}\left(\frac{H^2}{\mu}\frac{dp}{dx}\right)^2 = \frac{H^4}{k\mu}\left(\frac{dp}{dx}\right)^2.$$

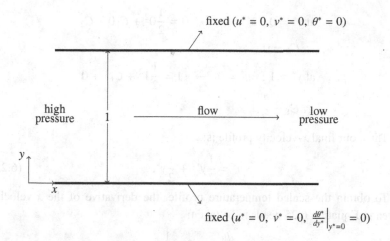

fixed ($u^* = 0$, $v^* = 0$, $\theta^* = 0$)

high
pressure

flow

low
pressure

fixed ($u^* = 0$, $v^* = 0$, $\left.\frac{d\theta^*}{dy^*}\right|_{y^*=0} = 0$)

Figure 6.6 The setup for pressure driven flow with scaled variables.

Thus our final scaled differential equations are much simpler:

$$\frac{d^2u^*}{dy^{*2}} = 1 \qquad (x\text{-momentum})$$

$$\frac{d^2\theta^*}{dy^{*2}} = -\left(\frac{du^*}{dy^*}\right)^2 \qquad (\text{energy}). \tag{6.23}$$

The scaled boundary conditions are:

$$\text{at } y^* = 0, \quad u = 0, \quad \text{and} \quad \left.\frac{d\theta^*}{dy^*}\right|_{y^*=0} = 0$$

$$\text{at } y^* = 1, \quad u = 0, \quad \text{and} \quad \theta^* = 0. \tag{6.24}$$

The scaled setup is illustrated in Figure 6.6.

Step 6: We can solve Equation 6.23 for the x-velocity and the temperature profiles. The general solution for the x-velocity is:

$$\frac{d^2u^*}{dy^{*2}} = 1 \xrightarrow{\text{solve}} u^* = \frac{1}{2}y^{*2} + C_1y^* + C_2.$$

Applying the boundary conditions in Equation 6.24 for the x-velocity leads to:

at $y^* = 0$, $u^* = 0$ $\rightarrow$ $0 = \frac{1}{2}0^2 + C_1 0 + C_2$

$\therefore C_2 = 0$

at $y^* = 1$: $u^* = 0$ $\rightarrow$ $1 = \frac{1}{2}1^2 + C_1 1 + 0$

$\therefore C_1 = \frac{1}{2}$.

Thus, our final x-velocity profile is:

$$u^* = \frac{1}{2}y^{*2} + \frac{1}{2}y^*. \tag{6.25}$$

To obtain the scaled temperature profile, the derivative of the x-velocity given in Equation 6.25 needs to be taken:

$$\frac{du^*}{dy^*} = y^* + \frac{1}{2}.$$

So, the general solution of the energy equation becomes (with the derivative of the x-velocity plugged in):

$$\frac{d^2\theta^*}{dy^{*2}} = -\left(\frac{du^*}{dy^*}\right)^2 \rightarrow \frac{d^2\theta^*}{dy^{*2}} = -\left(y^* + \frac{1}{2}\right)^2$$

$$\rightarrow \frac{d^2\theta^*}{dy^{*2}} = -\left(y^{*2} + y^* + \frac{1}{4}\right)$$

$$\xrightarrow{\text{solve}} \theta^* = -\frac{1}{12}y^{*4} - \frac{1}{6}y^{*3} - \frac{1}{8}y^{*2} + C_3 y^* + C_4.$$

We now need to apply the boundary conditions given in Equation 6.24:

at $y^* = 0$, $\frac{d\theta^*}{dy^*} = 0$ $\rightarrow$ $0 = -\frac{1}{3}0^3 - \frac{1}{2}0^2 - \frac{1}{4}0 + C_3$

$\therefore C_3 = 0$

at $y^* = 1$, $\theta^* = 0$ $\rightarrow$ $0 = -\frac{1}{12}1^4 - \frac{1}{6}1^3 - \frac{1}{8}1^2 + 0 \cdot 1 + C_4$

$\therefore C_4 = \frac{9}{24}$.

Therefore the final scaled temperature profile is:

$$\theta^* = -\frac{1}{12}y^{*4} - \frac{1}{6}y^{*3} - \frac{1}{8}y^{*2} + \frac{9}{24}. \tag{6.26}$$

The key thing to recognize here is that the nondimensional temperature and x-velocity now only depend on y^* with no other parameters involved. Therefore, with our current scaling, the scaled problem collapsed down to a single independent variable.

6.5 Scaling the Incompressible Governing Equations

Now, let's do a more general scaling for the incompressible flow equations (without heat generation). To do so, we are going to consider a scenario where the flow has a general length scale of L and a velocity scale of U_∞. The characteristic time scale, temperature scale, and pressure scale will not be defined right away but instead will be determined as we go.

The incompressible flow equations in dimensional form are (without heat generation):

Incompressible Continuity Equation:

$$\vec{\nabla} \cdot \vec{V} = 0$$

Incompressible Navier–Stokes Equations:

$$\rho\left(\frac{\partial \vec{V}}{\partial t} + \vec{V} \cdot \vec{\nabla}\vec{V}\right) = -\vec{\nabla}p + \mu\nabla^2\vec{V} + \rho\vec{g}$$

Incompressible Energy Equation (without Heat Generation)

$$\rho c_p\left(\frac{\partial T}{\partial t} + \vec{V} \cdot \vec{\nabla}T\right) = k\nabla^2 T + \mu\left(\vec{\nabla}\vec{V} + \left(\vec{\nabla}\vec{V}\right)^\dagger\right) : \vec{\nabla}\vec{V}.$$

We will assume the material properties (density, specific heat, thermal conductivity, and dynamic viscosity) are constant for this exercise. Note that in general this may not be the case. We are going to go through and scale each equation, starting with the continuity equation.

6.5.1 Scaled Continuity Equation for an Incompressible Flow

Scaling the continuity equation is quite simple, especially for an incompressible flow. The only variables we need to worry about are the velocity and the spatial derivatives (which are embedded in the nabla symbol). We can scale the velocity and nabla by using a velocity scale factor of U_∞ and a length scale of L. Thus:

$$\vec{\nabla}^* = \frac{\vec{\nabla}}{1/L}, \quad \vec{V}^* = \frac{\vec{V}}{U_\infty}.$$

Note that the length scale of the nabla symbol is actually the inverse of a length scale, L. The reason for this is because the nabla symbol has units (in SI) of $1/m$ because they contain spatial derivatives (i.e., $\frac{\partial}{\partial x}$, $\frac{\partial}{\partial y}$, etc.).

We can plug this scaling into the continuity equation for an incompressible flow to get:

$$\frac{U_\infty}{L}\vec{\nabla}^* \cdot \vec{V}^* = 0$$

$$\therefore \vec{\nabla}^* \cdot \vec{V}^* = 0.$$

Thus the **scaled continuity equation for an incompressible flow** is:

$$\boxed{\vec{\nabla}^* \cdot \vec{V}^* = 0}$$

Scaling the incompressible continuity equation yields an equation very similar to the unscaled equation. Note we divided out by $\frac{U_\infty}{L}$ in order to make the equation nondimensional.

6.5.2 Scaling the Incompressible Navier–Stokes Equations

Scaling the Navier–Stokes equations now requires some additional variables to scale, namely:

$$p^* = \frac{p}{p_s}, \quad t^* = \frac{t}{t_s}, \quad \vec{g}^* = \frac{\vec{g}}{g_0}$$

where p_s is a characteristic pressure scale, t_s is a characteristic time scale, and g_0 is a characteristic gravitational acceleration scale. We are going to assume that g_0 is a known scale (maybe it is just the value of the gravitational acceleration at sea level, 9.81 m/s^2). However, the characteristic pressure scale and the characteristic time scale will be determined as we proceed. Also notice that we are not scaling density and viscosity because we are assuming that they are constant.

We can introduce these scaled variables into the incompressible Navier–Stokes equations to get:

$$\rho\left(\frac{U_\infty}{t_s}\frac{\partial \vec{V}^*}{\partial t^*} + \frac{U_\infty^2}{L}\vec{V}^* \cdot \vec{\nabla}^*\vec{V}^*\right) = -\frac{p_s}{L}\vec{\nabla}^* p^* + \mu\frac{U_\infty}{L^2}\nabla^{*2}\vec{V}^* + \rho g_0\vec{g}^*.$$

We can pull out $\frac{U_\infty^2}{L}$ of the parentheses from the left-hand side of the equation to get:

$$\rho\frac{U_\infty^2}{L}\left(\underbrace{\frac{L}{U_\infty t_s}}_{\text{set}=1}\frac{\partial \vec{V}^*}{\partial t^*} + \vec{V}^* \cdot \vec{\nabla}^*\vec{V}^*\right) = -\frac{p_s}{L}\vec{\nabla}^* p^* + \mu\frac{U_\infty}{L^2}\nabla^{*2}\vec{V}^* + \rho g_0\vec{g}^*. \quad (6.27)$$

We can now define $t_s = L/U_\infty$ in order to obtain a term of 1 in front of $\frac{\partial \vec{V}}{\partial t}$ as is our standard procedure. Next, we need to decide which coefficient to divide out by: $\rho\frac{U_\infty^2}{L}$ on the left-hand side, $\frac{p_s}{L}$ in the pressure gradient term, $\mu\frac{U_\infty}{L^2}$ in the diffusion term, or the ρg_0 in the gravitational body force term.

Generally speaking, we have options for which one we divide out. So far, we have been dividing out by the left-most term. However, it is usually considered best practice to divide out by the term most dominant. For us, we are going to consider dividing out by two parameters: the diffusion parameter ($\mu \frac{U_\infty}{L^2}$) and the advection parameter ($\rho \frac{U_\infty^2}{L}$). We will start by dividing out the diffusion parameter.

Assuming the Diffusion Term is Dominant: Stokes Flow

Dividing out by the coefficient in the diffusive transport term, i.e., $\mu \frac{U_\infty}{L^2}$, in Equation 6.27 leads to:

$$\underbrace{\rho \frac{U_\infty L}{\mu}}_{Re} \left(\frac{\partial \vec{V}^*}{\partial t^*} + \vec{V}^* \cdot \vec{\nabla}^* \vec{V}^* \right) = -\underbrace{\frac{p_s L}{U_\infty \mu}}_{\text{set to 1}} \vec{\nabla}^* p^* + \nabla^{*2} \vec{V}^* + \frac{\rho g_0 L^2}{\mu U_\infty} \vec{g}^*. \quad (6.28)$$

Equation 6.28 is now a scaled, or nondimensional, equation. Note that the nondimensional parameter on the left-hand side of Equation 6.28 is Reynolds number. Also note that we are going to set the nondimensional coefficient in front of the pressure gradient to be one, so that now we can define the characteristic pressure scale for this scenario to be:

$$p_s = \frac{U_\infty \mu}{L}.$$

A very common simplification to Equation 6.28 is to assume that the Reynolds number approaches zero (since it is assumed to be diffusion dominant) and to ignore the gravitational body force. In doing so, Equation 6.28 becomes:

$$\boxed{0 = -\vec{\nabla}^* p^* + \nabla^{*2} \vec{V}^*} \quad (6.29)$$

Equation 6.29 is a very common simplification of the Navier–Stokes equations and many problems have been solved using Equation 6.29. The most famous of which is low Reynolds number flow past a sphere that was solved by George Stokes in 1851. In fact, the type of flow where Equation 6.29 is utilized (i.e., very low Reynolds number flow) is called **Stokes flow**, or sometimes **creeping flow**.

Assuming the Advection Term is Dominant: High Reynolds Number Flow

If we assume the advection term is dominant and divide Equation 6.27 out by $\rho \frac{U_\infty^2}{L}$ we get::

$$\left(\frac{\partial \vec{V}^*}{\partial t^*} + \vec{V}^* \cdot \vec{\nabla}^* \vec{V}^*\right) = - \underbrace{\frac{p_s}{\rho U_\infty^2}}_{\text{set} = 1} \vec{\nabla}^* p^* + \underbrace{\frac{\mu}{\rho U_\infty L}}_{\frac{1}{Re}} \nabla^{*2} \vec{V} + \frac{L g_0}{U_\infty^2} \vec{g}^*.$$

The characteristic pressure scale, p_s, is now set to be (in order to make the coefficient in front of the pressure gradient term equal to 1):

$$p_s = \rho U_\infty^2.$$

In addition, the coefficient in front of the diffusion term on the right-hand side turns out to be the inverse of Reynolds number. In addition, the coefficient in front of the gravity body force term can also be written in terms of a standard nondimensional parameter known as the Froude number (Fr). The Froude number is defined as:

$$Fr = \sqrt{\frac{U_\infty^2}{g_0 L}}.$$

Plugging in these common nondimensional numbers gives us the most common form for the **scaled incompressible Navier–Stokes equations**:

$$\boxed{\left(\frac{\partial \vec{V}^*}{\partial t^*} + \vec{V}^* \cdot \vec{\nabla}^* \vec{V}^*\right) = -\vec{\nabla}^* p^* + \frac{1}{Re} \nabla^{*2} \vec{V}^* + \frac{1}{Fr^2} \vec{g}^*} \qquad (6.30)$$

The key thing to note with Equation 6.30 is that the parameters of a problem no longer rely on dynamic viscosity, density, a length scale, and velocity scale. Instead, we have "collapsed" the number of parameters to two: Reynolds number and Froude number. Thus, you could have two completely different fluids moving at different velocities with differing length scales behaving the exact same way as long as the Reynolds and Froude number of the different scenarios are the same (in addition to the geometric shape of the problem being the same).

Considering this scenario assumed an advection-dominated flow, you might be tempted to look at what happens when Reynolds number approaches a very large number. Can we get rid of the diffusion term? The reality is that we may not always be able to eliminate the diffusion term. The reason why not will be discussed when we get into our boundary layer discussion a bit later. For now, just sit tight.

6.5.3 Scaling the Incompressible Energy Equation

To scale the incompressible energy equation, we are going to introduce the a scaled temperature whose characteristic scale factor is denoted by T_s and a general shift in temperature is given by T_{shift}, i.e.:

$$\theta^* = \frac{T - T_{shift}}{T_s}.$$

Introducing the above scaling for temperature and using the scaling for velocity, time, and the nabla operator from before, the energy equation becomes:

$$\rho c_p \left(\frac{U_\infty T_s}{L} \frac{\partial \theta^*}{\partial t^*} + \frac{U_\infty T_s}{L} \vec{V}^* \cdot \vec{\nabla}^* \theta^* \right) = k \frac{T_s}{L^2} \nabla^{*2} \theta^* + \mu \frac{U_\infty^2}{L^2} \left(\vec{\nabla}\vec{V} + \left(\vec{\nabla}\vec{V} \right)^\dagger \right) : \vec{\nabla}\vec{V}.$$

Recall that $t_s = \frac{L}{U_\infty}$ and in the viscous dissipation term we effectively have $\vec{\nabla}\vec{V}$ multiplied with itself. We are also ignoring heat generation. We can pull out the $\frac{U_\infty T_s}{L}$ term from the parentheses on the left-hand side to get:

$$\rho c_p \frac{U_\infty T_s}{L} \left(\frac{\partial \theta^*}{\partial t^*} + \vec{V}^* \cdot \vec{\nabla}^* \theta^* \right) = k \frac{T_s}{L^2} \nabla^{*2} \theta^* + \mu \frac{U_\infty^2}{L^2} \left(\vec{\nabla}\vec{V} + \left(\vec{\nabla}\vec{V} \right)^\dagger \right) : \vec{\nabla}\vec{V}.$$

In this situation, we are going to divide out by $\rho c_p \frac{U_\infty T_s}{L}$:

$$\frac{\partial \theta^*}{\partial t^*} + \vec{V}^* \cdot \vec{\nabla}^* \theta^* = \left(\frac{k}{\rho c_p U_\infty L} \nabla^{*2} \theta^* + \mu \frac{U_\infty}{\rho c_p T_s L} \left(\vec{\nabla}\vec{V} + \left(\vec{\nabla}\vec{V} \right)^\dagger \right) : \vec{\nabla}\vec{V} \right).$$

Interestingly enough, a very common thing to do next is to factor out a $\frac{1}{Re}$ $\left(\text{or } \frac{\mu}{\rho U_\infty L} \right)$ from all of the terms on the right-hand side, leaving us with:

$$\frac{\partial \theta^*}{\partial t^*} + \vec{V}^* \cdot \vec{\nabla}^* \theta^* = \frac{1}{Re} \left(\left(\frac{k}{c_p \mu} \right) \nabla^{*2} \theta^* + \frac{U_\infty^2}{c_p T_s} \left(\vec{\nabla}\vec{V} + \left(\vec{\nabla}\vec{V} \right)^\dagger \right) : \vec{\nabla}\vec{V} \right).$$

The nice thing about pulling out the Reynolds number, or more precisely the inverse of the Reynolds number, is that the coefficient in front of the diffusion term (i.e., the $\frac{k}{c_p \mu}$ term) is in terms of material properties and not flow properties. This ratio is the inverse of the Prandtl number that was defined earlier in Section 6.3: $Pr = \frac{c_p \mu}{k} = \frac{\nu}{\alpha}$.

Introducing the Pr number into the energy equation above leads to:

$$\frac{\partial \theta^*}{\partial t^*} + \vec{V}^* \cdot \vec{\nabla}^* \theta^* = \frac{1}{Re} \left(\left(\frac{1}{Pr} \right) \nabla^{*2} \theta^* + \frac{U_\infty^2}{c_p T_s} \left(\vec{\nabla}\vec{V} + \left(\vec{\nabla}\vec{V} \right)^\dagger \right) : \vec{\nabla}\vec{V} \right). \quad (6.31)$$

Next up is the viscous dissipation term. The coefficient in front of this term includes the characteristic temperature scale, T_s. We have some options for setting this characteristic temperature term, depending on the situation. Namely, if the problem has an obvious temperature scale or if the problem does not.

An Obvious Temperature Scale

An example of what one might consider an obvious temperature scale would be if the walls of your system have a fixed temperature, such as what was done in the example in Section 6.2.1. The temperature scale in this case would be something along the lines of a difference in the wall temperatures, e.g.:

$$T_s = T_{right} - T_{left} \quad \text{(or some other temperature difference, } \Delta T\text{)}.$$

Introducing this temperature scale into the energy equation would lead to:

$$\frac{\partial \theta^*}{\partial t^*} + \vec{V}^* \cdot \vec{\nabla}^* \theta^* = \frac{1}{Re}\left(\left(\frac{1}{Pr}\right)\nabla^{*2}\theta^* + \frac{U_\infty^2}{c_p \Delta T}\left(\vec{\nabla}\vec{V} + \left(\vec{\nabla}\vec{V}\right)^\dagger\right):\vec{\nabla}\vec{V}\right).$$

The coefficient next to the viscous dissipation terms is something we have seen before in Section 6.3, it is the Eckert number (Ec): $Ec = \frac{U_\infty^2}{c_p \Delta T}$.

Introducing the Eckert number into the energy equation gives us:

$$\frac{\partial \theta^*}{\partial t^*} + \vec{V}^* \cdot \vec{\nabla}^* \theta^* = \frac{1}{Re}\left(\left(\frac{1}{Pr}\right)\nabla^{*2}\theta^* + Ec\left(\vec{\nabla}\vec{V} + \left(\vec{\nabla}\vec{V}\right)^\dagger\right):\vec{\nabla}\vec{V}\right) \quad (6.32)$$

Multiplying the Reynolds number through and recalling the Péclet number is defined as: $Pe = \frac{U_\infty L}{\alpha}$, which is just the same as $Pe = RePr = \frac{U_\infty L}{\nu}\frac{\nu}{\alpha} = \frac{U_\infty L}{\alpha}$, would give:

$$\frac{\partial \theta^*}{\partial t^*} + \vec{V}^* \cdot \vec{\nabla}^* \theta^* = \underbrace{\frac{1}{Pe}}_{=\frac{1}{RePr}}\nabla^{*2}\theta^* + \frac{Ec}{Re}\left(\vec{\nabla}\vec{V} + \left(\vec{\nabla}\vec{V}\right)^\dagger\right):\vec{\nabla}\vec{V} \quad (6.33)$$

Equations 6.32 and 6.33 are examples of the **scaled energy equation for an incompressible flow with a temperature difference scale**. Notice that when the characteristic velocity is really small, i.e., as U_∞ approaches 0, the Eckert number approaches zero more quickly than the Reynolds number (since the Eckert number contains a velocity squared term). Thus, in really slow-moving flows, the viscous dissipation term can often be neglected because the Eckert number often approaches zero much quicker than the Reynolds number.

Not an Obvious Temperature Scale

We now come to the situation where there is not an obvious temperature scale, such as the situation we came across with an insulated wall in Section 6.4. In such a situation, we often set the characteristic temperature scale such that the term in front of the viscous dissipation is one, i.e., Equation 6.31 becomes:

$$\frac{\partial \theta^*}{\partial t^*} + \vec{V}^* \cdot \vec{\nabla}^* \theta^* = \frac{1}{Re}\left(\left(\frac{1}{Pr}\right)\nabla^{*2}\theta^* + \underbrace{\frac{U_\infty^2}{c_p T_s}}_{\text{set}=1}\left(\vec{\nabla}\vec{V} + \left(\vec{\nabla}\vec{V}\right)^\dagger\right):\vec{\nabla}\vec{V}\right).$$

Therefore the T_s is:

$$T_s = \frac{U_\infty^2}{c_p}.$$

This makes the energy equation become the **scaled energy equation for an incompressible flow with insulated walls**:

$$\frac{\partial \Theta^*}{\partial t^*} + \vec{V}^* \cdot \vec{\nabla}^* \Theta^* = \frac{1}{Re}\left(\left(\frac{1}{Pr}\right)\nabla^{*2}\Theta^* + \left(\vec{\nabla}\vec{V} + \left(\vec{\nabla}\vec{V}\right)^\dagger\right) : \vec{\nabla}\vec{V}\right) \quad (6.34)$$

Equations 6.32, 6.33, and 6.34 are all typical options for the scaled energy equation. As is typical with the nondimensionalization of a differential equation, scaling the energy equation reduces the number of parameters in the problem. The parameters in the dimensional problem would consist of the size (i.e., length scale) of the system, the values of the temperature or heat flux at the boundaries, the velocity information at the boundaries, the density, the dynamic viscosity, the specific heat, and the thermal conductivity. In the scaled problem, the energy equation (and its boundary conditions) collapse down so that you normally would only have to compute the Reynolds number, the Prandtl number, and the Eckert number (or Reynolds, Péclet, and Eckert number). You might be wondering why you wouldn't just always use Equation 6.34 since the number of parameters is even smaller (only Reynolds and Prandtl number), and the reason has to do with the boundary conditions. In some problems, having a characteristic temperature scale associated with a temperature difference will simplify the boundary conditions greatly.

6.6 Incompressible Flow with a Compressible Fluid

A compressible fluid, such as air, is one where the density is easily changed with a change in pressure. In an incompressible fluid, such as liquid water under normal circumstances, the density does not change significantly with pressure.[2] As a result, flow situations involving liquid water are typically modeled as incompressible flows (at least in engineering applications) because the overall density change is usually not that significant. However, in many situations involving air (such as in certain atmospheric studies or in the design of aircraft), you would not be able to use the incompressible flow equations since the density of the airflow easily changes with pressure. In addition, such compressible fluids also usually obey equations of state where the density, pressure,

---

[2] It should be noted that all fluids have some level of compressibility.

and temperature are all highly dependent on each other. This begs the question: is there ever a situation where the flow of a compressible fluid, such as air, can be modeled using the incompressible governing equations? We can investigate this question using scaling analysis, mixed with a little bit of thermodynamics.

To begin the investigation, consider a fluid whose density can readily change either via temperature or pressure. In other words, density is a function of pressure and temperature, i.e., $\rho = \rho(p, T)$, much like in an ideal gas the density is very dependent on the pressure and temperature via $\rho = \frac{p}{RT}$.

Taking the material derivative of density (you will see why this is done shortly), we get the following expression:

$$\frac{D\rho}{Dt} = \frac{D}{Dt}(\rho(p, T))$$
$$= \left(\frac{\partial\rho}{\partial p}\right)_T \frac{Dp}{Dt} + \left(\frac{\partial\rho}{\partial T}\right)_p \frac{DT}{Dt} \quad \text{(chain rule).}$$

(6.35)

Recall from Chapter 5 that the isothermal compressibility is defined as: $\beta_p = \frac{1}{\rho}\left(\frac{\partial\rho}{\partial p}\right)_T$ and the coefficient of thermal expansion is: $\beta_T = -\frac{1}{\rho}\left(\frac{\partial\rho}{\partial T}\right)_p$. Using this information, we can now write Equation 6.35 as:

$$\frac{D\rho}{Dt} = \rho\beta_p \frac{Dp}{Dt} - \rho\beta_T \frac{DT}{Dt}.$$

(6.36)

The scaling we will use for this equation is going to be:

$$\rho^* = \frac{\rho}{\rho_0}, \quad p^* = \frac{p}{\rho_0 U_\infty^2}, \quad \theta^* = \frac{T - T_{shift}}{U_\infty^2 / c_{p0}}, \quad t^* = \frac{t}{t_s},$$

$$\beta_p^* = \frac{\beta_p}{\beta_{p0}}, \quad \beta_T^* = \frac{\beta_T}{\beta_{T0}}.$$

Notice that we are no longer assuming a constant density, hence the reason we are including it in the scaling. The 0 subscripts denote a value that is known while the s subscript indicates that the characteristic scale is to be determined (although, as you will see, t_s will just divide out and we won't have to worry about it). Also, notice that the scaling we are using for temperature is the scaling we used for insulated walls. This is merely a convenience as the result we are after is more easily shown with this scaling.

Introducing the scaling into Equation 6.36 leads to:

$$\frac{\rho_0}{t_s}\frac{D\rho^*}{Dt^*} = \rho_0\beta_{p0}\frac{\rho_0 U_\infty^2}{t_s}\rho^*\beta_p^*\frac{Dp^*}{Dt^*} - \rho_0\beta_{T0}\frac{U_\infty^2/c_{p0}}{t_s}\rho^*\beta_T^*\frac{D\theta^*}{Dt^*}.$$

Dividing out by the $\frac{\rho_0}{t_s}$ coefficient on the left-hand side yields:

$$\frac{D\rho^*}{Dt^*} = \rho_0\beta_{p0}U_\infty^2\rho^*\beta_p^*\frac{Dp^*}{Dt^*} - \beta_{T0}\frac{U_\infty^2}{c_{p0}}\rho^*\beta_T^*\frac{D\theta^*}{Dt^*}.$$

(6.37)

From thermodynamics, the speed of sound in a fluid is defined as:[3]

$$a_o^2 = \frac{c_{p0}/c_{v0}}{\rho_0 \beta_{p0}}$$

where c_{p0} and c_{vo} are the characteristic scales for the specific heat values at constant pressure and volume, respectively. The speed of sound deals with the speed of propagation of small pressure waves (i.e., sound waves) in a medium.

Replacing the $\rho_0 \beta_{p0}$ with $\frac{c_{p0}/c_{v0}}{a_o^2}$ in the first term on the right and c_{p0} with $a_o^2 \rho_0 \beta_{p0} c_{v0}$ in the second term on the right in Equation 6.37 leads to:

$$\frac{Dp^*}{Dt^*} = \frac{c_{p0}/c_{v0}}{a_o^2} U_\infty^2 \rho^* \beta_p^* \frac{Dp^*}{Dt^*} - \beta_{T0} \frac{U_\infty^2}{a_o^2 \rho_0 \beta_{p0} c_{v0}} \rho^* \beta_T^* \frac{D\theta^*}{Dt^*}.$$

We can rearrange slightly to get:

$$\frac{Dp^*}{Dt^*} = \frac{c_{p0}}{c_{v0}} \frac{U_\infty^2}{a_o^2} \rho^* \beta_p^* \frac{Dp^*}{Dt^*} - \frac{\beta_{T0}}{\rho \beta_{p0} c_{v0}} \frac{U_\infty^2}{a_o^2} \rho^* \beta_T^* \frac{D\theta^*}{Dt^*}. \tag{6.38}$$

We are now in a position to introduce what is known as the Mach number, named after Ernst Mach in the late eighteen hundreds. The Mach number is defined as the velocity of the flow divided by the speed of sound, i.e.:

$$Ma = \frac{U_\infty}{a_o}.$$

Introducing the Mach number into Equation 6.38 gives us:

$$\frac{Dp^*}{Dt^*} = \frac{c_{p0}}{c_{v0}} Ma^2 \rho^* \beta_p^* \frac{Dp^*}{Dt^*} - \frac{\beta_{T0}}{\rho \beta_{p0} c_{v0}} Ma^2 \rho^* \beta_T^* \frac{D\theta^*}{Dt^*}. \tag{6.39}$$

Notice what happens when the Mach number goes to zero in Equation 6.39, i.e., when $Ma \to 0$. The right-hand side goes to zero and we are left with:

$$\frac{Dp^*}{Dt^*} = 0.$$

You may recall this indicates that the density of a fluid element (i.e., the volume of a fluid element) does not change as it moves, which is the definition of an incompressible flow. The continuity equation then results in the typical $\vec{\nabla} \cdot \vec{V} = 0$ as seen in incompressible flow situations. Thus, if a compressible fluid (such as air) is moving slow enough, it can still be considered an incompressible flow. As such, we now have come to our last definition of an incompressible flow:

---

[3] Showing how the speed of sound comes about from thermodynamics would, admittedly, deviate from the discussion.

An incompressible flow is one where the Mach number approaches zero.

A standard "cut-off" value of Mach number in order to be considered incompressible is typically around 0.3 for engineering applications. That is, a flow is generally considered to be incompressible when the maximum flow velocity is such that the Mach number (using the max velocity) is lower than 0.3. Any flow that has a velocity such that the Mach number is greater than 0.3 is usually considered a compressible flow.

6.7 Scaling to Obtain the Boundary Layer Equations

Let's take another look at the incompressible Navier–Stokes equations. The general scaled form of the incompressible Navier–Stokes equations (ignoring the gravitational body force and not bothering to write the continuity equation) is Equation 6.30:

$$\frac{\partial \vec{V}^*}{\partial t^*} + \vec{V}^* \cdot \vec{\nabla}^* \vec{V}^* = -\vec{\nabla}^* p^* + \frac{1}{Re}\nabla^{*2}\vec{V}^*.$$

If we take the Reynolds number to infinity, the scaled Navier–Stokes equations become:

$$\frac{\partial \vec{V}^*}{\partial t^*} + \vec{V}^* \cdot \vec{\nabla}^* \vec{V}^* = -\vec{\nabla}^* p^* + \overset{0}{\cancel{\frac{1}{Re}}}\nabla^{*2}\vec{V}^*$$

$$\rightarrow \frac{\partial \vec{V}^*}{\partial t^*} + \vec{V}^* \cdot \vec{\nabla}^* \vec{V}^* = -\vec{\nabla}^* p^*. \tag{6.40}$$

Equation 6.40 is the scaled Euler equation. We introduced the Euler equation in Chapter 3. It governs inviscid flow (i.e., no viscosity) and is easier to handle than the Navier–Stokes equations. The Euler equation does not have a diffusive transport (i.e., a viscous force) term. There is a whole subset of solutions for inviscid flow known as potential flow. There are issues, however, with using the Euler equation and the potential flow solutions. The biggest issue is that it does not provide accurate drag values (i.e., the force parallel to the flow that is exerted on an object such as a sphere or flat plate). In particular, the solutions often give drag values of zero, even though this is known to not be the case. The contradiction between the results of the analysis assuming inviscid flow and the experimental evidence is called the D'Alembert's paradox. Part of the issue has to do with the number of boundary conditions needed for the Navier–Stokes equations versus the Euler equations. Recall that the number of boundary conditions in a given direction is usually related to the highest order derivative in a

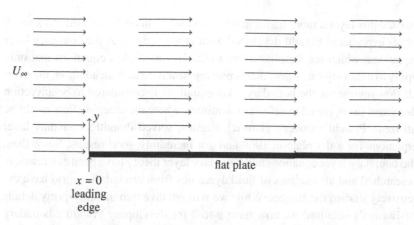

Figure 6.7 The setup for flow past a flat plate assuming no diffusion term (i.e., no viscosity).

differential equation. The Euler equation has only first-order derivatives. As a result, the Euler equation does not need as much boundary information since it is a first order, as opposed to second order, differential equation. A simple way to understand why this is a problem is to consider the classic example of flow past a plate. If we were to solve for the velocity profile in flow past a flat plate using Euler's equations, we really only need to satisfy a boundary condition at the far stream away from the plate, i.e.: at $y \to \infty$, $u \to U_\infty$, where U_∞ is the free stream velocity. The solution would end up looking much like what is seen in Figure 6.7. Notice in Figure 6.7 that the velocity vectors do not change as they approach the plate. We know that the velocity of the flow has to slow down near the plate due to the friction (or diffusion) at the plate, and this friction is what causes the drag on the plate. In order to force the velocity to slow down at the plate, a second boundary condition is needed at the plate. That boundary condition is the no-slip boundary condition. However, in order to be able to apply a second boundary condition, the second order derivative term (i.e., the diffusion term) needs to still be included (no matter how big Reynolds is and thus, how small the coefficient $\frac{1}{Re}$ is in front of the second order derivatives). Mathematically, this problem is called a singular perturbation problem. Singular perturbation problems occur in differential equations when there is a very small term in front of the highest-order derivatives.

So if the diffusion term needs to be included, how can we make any simplifications to the Navier–Stokes equations for this problem? The answer came from a German physicist named Ludwig Prandtl. Prandtl hypothesized in 1904 that there was a small region near the boundary (even in very high speed or high Reynolds number flows) where the viscosity is most important, and outside this layer a more traditional inviscid solution could be used. As a result of his hypothesis, Prandtl developed what is now known as the boundary layer equations, which are simplifications to the Navier–Stokes equations that only apply to a thin region of flow that is passing near a surface. In addition, the overall flow regime for the boundary layer equations is considered to be advection dominant (as opposed to diffusion dominant where the creeping flow might be utilized). Prandtl's student, Heinrich Blasius, solved Prandtl's boundary layer equations for a flat plate in 1908 and got incredibly good results. Since then, the boundary layer equations and boundary layer theory have been exhaustively researched and all students of fluid dynamics from around the world have extensively studied the subject. While we will not dive into the nitty-gritty details of Blasius's solution, we now have a tool for developing Prandtl's boundary layer equations: scaling.

We start off by writing down the two-dimensional incompressible continuity and Navier–Stokes equations (Equations 4.24, 4.25, and 4.26). We are not going to bother with the energy equation. In addition, we are going to assume a steady state scenario (i.e., the time derivatives go to zero) as well as ignore any gravitational effects. Thus, our current simplified set of equations looks like this:

$$\frac{\partial u}{\partial x} + \frac{\partial v}{\partial y} = 0. \tag{6.41}$$

$$\rho\left(u\frac{\partial u}{\partial x} + v\frac{\partial u}{\partial y}\right) = -\frac{\partial p}{\partial x} + \mu\left(\frac{\partial^2 u}{\partial x^2} + \frac{\partial^2 u}{\partial y^2}\right). \tag{6.42}$$

$$\rho\left(u\frac{\partial v}{\partial x} + v\frac{\partial v}{\partial y}\right) = -\frac{\partial p}{\partial y} + \mu\left(\frac{\partial^2 v}{\partial x^2} + \frac{\partial^2 v}{\partial y^2}\right). \tag{6.43}$$

We introduce nondimensional (i.e., scaled) variables:

$$u^* = \frac{u}{U_\infty}, \quad v^* = \frac{v}{V_s}, \quad p^* = \frac{p}{p_s}, \quad x^* = \frac{x}{X}, \quad y^* = \frac{y}{\delta} \tag{6.44}$$

where U_∞ is the free stream velocity, V_s is a "to be determined" characteristic scale for the y-velocity, p_s is a "to be determined" characteristic pressure scale, X is some position downstream from the leading edge, and δ is the approximate thickness of the boundary layer at location X. The approximate boundary layer thickness, δ, and the distance downstream, X, are inter-related since the boundary layer thickness can change depending on how far downstream we are on the plate. This is depicted in Figure 6.8.

Plugging in our scaled variables into the continuity equation yields:

$$\frac{U_\infty}{X}\frac{\partial u^*}{\partial x^*} + \frac{V_s}{\delta}\frac{\partial v^*}{\partial y^*} = 0.$$

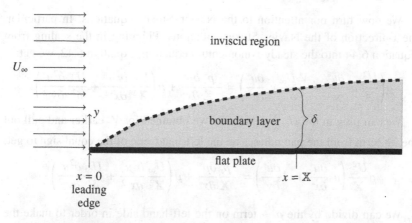

Figure 6.8 Illustration of a boundary layer in flow past a flat plate. The dashed line is the start of the boundary layer. Outside the boundary layer is an inviscid region and in the grey area is a viscous boundary layer region. The boundary layer thickness increases in size the further downstream from the leading edge of the plate.

The above equation has units of inverse seconds. To nondimensionalize, we will divide by the largest order term, $\frac{U_\infty}{X}$, in order to get:

$$\frac{\partial u^*}{\partial x^*} + \underbrace{\frac{V_s X}{\delta U_\infty}}_{\text{set to } 1} \frac{\partial v^*}{\partial y^*} = 0. \qquad (6.45)$$

Notice we have a three variables whose values are not fixed at this point (X, V_s, and δ). However, we are going to make an assumption that the weight of the two terms in the continuity equation given in Equation 6.45 will be the same. Thus, we will set the coefficient for the second term to one. Doing so allows us to obtain a relationship between the velocity scale in the y-direction and the velocity scale in the x-direction:

$$V_s = U_\infty \frac{\delta}{X}.$$

This relationship indicates that the velocity scale in the y-direction will probably be a fair amount smaller since the boundary layer thickness (δ) will generally be quite a bit smaller than the downstream distance (X). Inserting the relationship for V_s above into Equation 6.45, we get the scaled continuity equation:

$$\frac{\partial u^*}{\partial x^*} + \frac{\partial v^*}{\partial y^*} = 0. \qquad (6.46)$$

We now turn our attention to the Navier–Stokes equations. In particular, the x-direction of the Navier–Stokes equations. Plugging in the scaling from Equation 6.44 into the steady x-momentum equation, Equation 6.42, we get:

$$\rho\left(\frac{U_\infty^2}{X}u^*\frac{\partial u^*}{\partial x^*} + \frac{V_s U_\infty}{\delta}v^*\frac{\partial u^*}{\partial y^*}\right) = -\frac{p_s}{X}\frac{\partial p^*}{\partial x^*} + \mu\left(\frac{U_\infty}{X^2}\frac{\partial^2 u^*}{\partial x^{*2}} + \frac{U_\infty}{\delta^2}\frac{\partial^2 u^*}{\partial y^{*2}}\right).$$

We can plug in the $U_\infty\frac{\delta}{X}$ expression we obtained for V_s earlier and pull out the $\frac{U_\infty^2}{X}$ term from the parentheses on the left-hand side of the equal sign to get:

$$\rho\frac{U_\infty^2}{X}\left(u^*\frac{\partial u^*}{\partial x^*} + v^*\frac{\partial u^*}{\partial y^*}\right) = -\frac{p_s}{X}\frac{\partial p^*}{\partial x^*} + \mu\left(\frac{U_\infty}{X^2}\frac{\partial^2 u^*}{\partial x^{*2}} + \frac{U_\infty}{\delta^2}\frac{\partial^2 u^*}{\partial y^{*2}}\right).$$

We can divide by the $\rho\frac{U_\infty^2}{X}$ term on the left-hand side in order to make the equation nondimensional:

$$u^*\frac{\partial u^*}{\partial x^*} + v^*\frac{\partial u^*}{\partial y^*} = -\underbrace{\frac{p_s}{\rho U_\infty^2}}_{\text{set to 1}}\frac{\partial p^*}{\partial x^*} + \mu\frac{X}{U_\infty^2}\left(\frac{U_\infty}{X^2}\frac{\partial^2 u^*}{\partial x^{*2}} + \frac{U_\infty}{\delta^2}\frac{\partial^2 u^*}{\partial y^{*2}}\right). \qquad (6.47)$$

The x-momentum is now nondimensional. Setting the characteristic pressure scale to be $p_s = \rho U_\infty^2$ as well as factoring out the $\frac{U_\infty}{\delta^2}$ from the diffusion terms, makes Equation 6.47 become:

$$u^*\frac{\partial u^*}{\partial x^*} + v^*\frac{\partial u^*}{\partial y^*} = -\frac{\partial p^*}{\partial x^*} + \mu\frac{X}{\rho U_\infty^2}\frac{U_\infty}{\delta^2}\left(\underbrace{\frac{\delta^2}{X^2}}_{\text{small}}\frac{\partial^2 u^*}{\partial x^{*2}} + \frac{\partial^2 u^*}{\partial y^{*2}}\right).$$

The next thing we can do is take a look at the diffusive transport term. We are going to make yet another assumption, that the boundary layer thickness is much smaller than the distance downstream from the leading edge, or $\delta \ll X$. Doing so means that the diffusion term in the x-direction will be negligible since $\frac{\delta^2}{X^2} \ll 1$, leaving the following:

$$u^*\frac{\partial u^*}{\partial x^*} + v^*\frac{\partial u^*}{\partial y^*} = -\frac{\partial p^*}{\partial x^*} + \underbrace{\mu\frac{X}{\rho U_\infty^2}\frac{U_\infty}{\delta^2}}_{\text{set to 1}}\left(\frac{\partial^2 u^*}{\partial y^{*2}}\right). \qquad (6.48)$$

The start of the boundary layer occurs at the point where the diffusive transport and the advective transport terms are on equal footing, meaning we can make the $\mu\frac{X}{U_\infty^2}\frac{U_\infty}{\delta^2}$ coefficient in front of the diffusion term equal to one. This implies that the boundary layer thickness has the following relationship:

$$\delta = \sqrt{\mu\frac{X U_\infty}{\rho U_\infty^2}} = \sqrt{\frac{\mu X}{\rho U_\infty}}.$$

If we multiply the above equation by "one" via $\sqrt{\frac{X}{X}}$, we get:

$$\delta = \sqrt{\frac{\mu X}{\rho U_\infty}} \sqrt{\frac{X}{X}} = \sqrt{\frac{\mu}{\rho U_\infty X}} X.$$

We can introduce a Reynolds number whose length scale depends on the distance X downstream from the leading edge (Re_x), i.e.:

$$Re_x = \frac{\rho U_\infty X}{\mu}.$$

Thus, an estimate for the boundary layer thickness as a function of Reynolds number turns out to be:

$$\delta = \frac{X}{\sqrt{Re_x}}. \tag{6.49}$$

This is a hugely useful result as it provides, just on the basis of pure scaling arguments, a relationship between the boundary layer thickness and the Reynolds number. This is the same estimate we achieved in Chapter 5 when we discussed flow past a flat plate. The key finding is that the higher the Reynolds number, the thinner the boundary layer at a given X value. The reason for this is due to the fact that at a higher Reynolds number (such as a case with a high flow velocity), there will be more of a "push" from the high-velocity flow and thus the friction provided by the diffusive transport term will not extend as far into the flow field. Thus, in general, the boundary layer will be thinner.

Going back to our scaled x-momentum equation, with the relationship for the boundary layer thickness now defined, the x-momentum equation for the boundary layer is:

$$u^* \frac{\partial u^*}{\partial x^*} + v^* \frac{\partial u^*}{\partial y^*} = -\frac{\partial p^*}{\partial x^*} + \frac{\partial^2 u^*}{\partial y^{*2}}. \tag{6.50}$$

We can now take a look at the y-momentum. With the introduction of the scaled variables (Equation 6.44) into the steady y-momentum equation (Equation 6.43) becomes:

$$\rho \left(\frac{U_\infty V_s}{X} u^* \frac{\partial v^*}{\partial x^*} + \frac{V_s^2}{\delta} v^* \frac{\partial v^*}{\partial y^*} \right) = -\frac{p_s}{\delta} \frac{\partial p^*}{\partial y^*} + \mu \left(\frac{V_s}{X^2} \frac{\partial^2 v^*}{\partial x^{*2}} + \frac{V_s}{\delta^2} \frac{\partial^2 v^*}{\partial y^{*2}} \right).$$

We can pull out $\frac{V_s}{\delta^2}$ from the parenthesis in the diffusion term on the right-hand side to get:

$$\rho \left(\frac{U_\infty V_s}{X} u^* \frac{\partial v^*}{\partial x^*} + \frac{V_s^2}{\delta} v^* \frac{\partial v^*}{\partial y^*} \right) = -\frac{p_s}{\delta} \frac{\partial p^*}{\partial y^*} + \mu \frac{V_s}{\delta^2} \left(\frac{\delta^2}{X^2} \frac{\partial^2 v^*}{\partial x^{*2}} + \frac{\partial^2 v^*}{\partial y^{*2}} \right).$$

Notice that the first coefficient in the diffusion term on the right-hand side (i.e., the $\frac{\delta^2}{X^2}$ coefficient) becomes very small, so we will get rid of that term. In addition, we can introduce the scale factors for the y-velocity ($V_s = \frac{U_\infty \delta}{X}$) and the pressure ($p_s = \rho U_\infty^2$) to get:

$$\rho\left(\frac{U_\infty^2 \delta}{X^2}u^*\frac{\partial v^*}{\partial x^*} + \frac{U_\infty^2 \delta}{X^2}v^*\frac{\partial v^*}{\partial y^*}\right) = -\frac{\rho U_\infty^2}{\delta}\frac{\partial p^*}{\partial y^*} + \mu\frac{U_\infty}{\delta X}\left(\frac{\partial^2 v^*}{\partial y^{*2}}\right).$$

This equation is still not nondimensionalized. To do so, we will need to divide out by one of the coefficients containing the characteristic scale factors. This time we are going to divide out by the coefficient in front of the pressure gradient term (i.e., the $\frac{\rho U_\infty^2}{\delta}$ coefficient) since it appears to be of the largest magnitude (due to the fact that it is divided by the smaller δ). Doing so will lead to:

$$\underbrace{\left(\frac{\delta^2}{X^2}u^*\frac{\partial v^*}{\partial x^*} + \frac{\delta^2}{X^2}v^*\frac{\partial v^*}{\partial y^*}\right.}_{\text{goes to zero because } \frac{\delta^2}{X^2} \text{ is small}} = -\frac{\partial p^*}{\partial y^*} + \underbrace{\frac{\mu}{\rho U_\infty X}}_{\frac{1}{Re_x}<<1}\left.\left(\frac{\partial^2 v^*}{\partial y^{*2}}\right).\right.$$

The left-hand side will go to zero because of the small $\frac{\delta^2}{X^2}$ factor in front of both terms. The diffusion term on the right-hand side now has a coefficient of $\frac{1}{Re_x}$. For large Reynolds numbers, this term will have a minimal impact compared to the pressure gradient term. Thus, the final expression for the y-momentum is nothing but the pressure gradient in the y-direction equals zero:

$$\frac{\partial p^*}{\partial y^*} = 0. \tag{6.51}$$

Interestingly enough, this equation implies that the pressure only depends on x and not on y. Thus, the pressure will be the same whether you are inside the boundary layer or outside the boundary layer at a given x. So we can replace the pressure gradient in Equation 6.50 (which is the x-momentum equation inside the boundary layer) with a pressure gradient outside the boundary layer. So, what is the pressure gradient outside the boundary layer? In our scenario, the velocity of the flow does not change outside the boundary layer and is a uniform and constant U_∞. Given this fact, there can be no pressure gradient since the inclusion of a pressure gradient would indicate a net force acting on the fluid outside the boundary layer. Such a net force would cause the flow to either accelerate or decelerate due to Newton's laws of motion. Thus, the pressure derivative in the x-direction must be zero:

$$\frac{\partial p}{\partial x} = 0.$$

We now have a simplified set of equations for this problem, written here in scaled form:

Continuity for an incompressible flow past a flat plate:

$$\frac{\partial u^*}{\partial x^*} + \frac{\partial v^*}{\partial y^*} = 0$$

x-momentum for a steady incompressible flow past a flat plate (with a constant free stream velocity):

$$u^* \frac{\partial u^*}{\partial x^*} + v^* \frac{\partial u^*}{\partial y^*} = \frac{\partial^2 u^*}{\partial y^{*2}}$$

y-momentum for a steady incompressible flow past a flat plate:

$$\frac{\partial p^*}{\partial y^*} = 0.$$

These equations are considered the boundary layer equations for flow past a flat plate where the free stream velocity is a constant. As mentioned, the equations were developed by Ludwig Prandtl in 1904, and his student, Heinrich Blasius, solved them in 1908. The method Blasius used involved what is called a similarity transformation to transform the partial differential equations into a single non-linear ordinary differential equation. We will not do this here as it is somewhat involved. The velocity profile he developed matched experiments. A typical plot of the velocity profile within the boundary layer is given in Figure 6.9. With a velocity profile now available, a more precise definition for the boundary layer thickness can be utilized. The standard definition for the boundary layer thickness is the 'y-point location where the velocity of the flow is 99 percent of the free stream velocity. Thus, the more precise definition of the boundary layer thickness, δ_{BL}, is evaluated via:

$$\delta_{BL} = y\Big|_{u=0.99U_\infty} \quad \xrightarrow[\text{scaled variables}]{\text{or using}} \quad \delta_{BL} = \delta y^*\Big|_{u^*=0.99}.$$

The boundary layer thickness is illustrated in Figure 6.9. The criterion used to calculate the boundary layer thickness is just an agreed-upon point. Using the velocity profile obtained by Blasius's solution to Prandtl's boundary layer equations, the value of $y^*\Big|_{u^*=0.99} = 5$, thus the boundary layer thickness can be found to be:

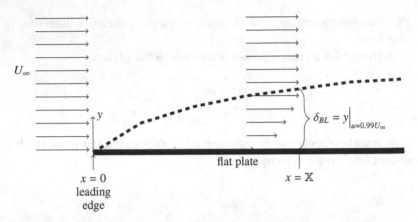

Figure 6.9 The velocity profile in a boundary layer. The thickness of the boundary layer is defined as the y-point where the velocity is $0.99U_\infty$ at a given $\mathbb{X}$.

$$\delta_{BL} = 5 \underbrace{\frac{\mathbb{X}}{\sqrt{Re_x}}}_{\delta}.$$

Some comments:

- This analysis only applies when the Reynolds number is greater than one. If the Reynolds number is less than one, the problem is diffusion dominant and there is no boundary layer.
- The relationship for the boundary layer thickness given above breaks down in turbulent flows. Turbulent flows, which the Navier–Stokes equations still govern, are too chaotic and unpredictable. Figure 6.10 illustrates the boundary layer for various regions of the flow, including laminar (which is what we have been dealing with), transition (not quite turbulent yet), and turbulent flow. The flow becomes turbulent as it continues to move downstream. The point at which the flow becomes turbulent is not always obvious. In general, however, engineers and physicists consider flow past a flat plate transitioning into turbulence when the Reynolds number is $Re_x = 500000$.

6.8 A Final Note

This book just scratched the surface of the Navier–Stokes equations. There are also other approaches to looking at the Navier–Stokes equations that we have

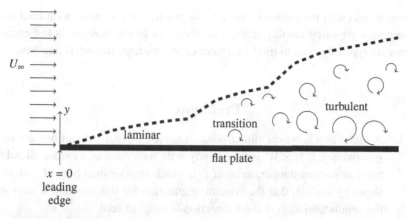

Figure 6.10 The setup for flow past a flat plate in the case where the flow transitions to turbulent flow downstream. Note the transition phase usually does not span a great distance.

not touched on. In particular, the Navier–Stokes equations can be transformed into an equation dealing with what is called vorticity. Vorticity is a flow variable that is often associated with the swirls (or vortices) of a fluid in motion. Using vorticity to study fluid mechanics, as opposed to velocity and pressure, has its advantages. It is particularly useful in the study of turbulence. As alluded to, turbulence deals with the chaotic nature of fluid motion. Turbulence is seemingly unpredictable and often shows up in relatively high Reynolds number situations. In other words, turbulence shows up in advection-dominated (or inertia-dominated) flows. In such flows, the advection effects attempt to push the flow in a certain direction while still being impeded by viscous forces, causing a large mess of vortices to begin to show up in unpredictable ways. It should be noted that turbulence does not just mean flow with vortices. Laminar flow can contain vortices as long as the vortices show up in a predictable and regular pattern. Nevertheless, even though the defining feature of turbulence is its unpredictability, the Navier–Stokes equations (whether in the forms we discussed in this book or in its vorticity form) are believed to apply in such situations. In computational fluid dynamics studies, where the Navier–Stokes equations are solved on a computer, the Navier–Stokes equations have been used to model turbulent flow. However, as the Reynolds number increases, the computer programs used to solve the Navier–Stokes equations need a lot more computer time to crunch the numbers because turbulent flow spans a large range of length and time scales (this is especially true for very high Reynolds numbers). Capturing all of the information in a very high Reynolds number flow in a computer program can start to become prohibitive in practice. As a result, additional assumptions are often employed and modifications to the Navier–Stokes equations are typically used to model turbulence at very high Reynolds numbers.

Problems

6.1   A metal block, whose dimensions in the x-, y-, and z-directions are respectively $L \times H \times W$, is in a steady state with each of its sides all held fixed at various temperatures. If L is much smaller than both H and W, show, by scaling, that the governing equation for this problem is simply the conduction term in the x-direction is equal to zero.

6.2   Scale the one-dimensional heat equation with the heat generation term included. Write the nondimensional heat equation in terms of Fourier number (Fo), a nondimensional time parameter defined as:

$$Fo = \frac{\alpha t}{L^2}$$

where α is the thermal diffusivity and L is the length scale.

6.3   Consider a conduction problem similar to the one given in Figure 5.5, except the boundary condition of the right side (at $x = L$) is no longer held at a fixed temperature and is instead in contact with a fluid at temperature T_∞. The boundary condition on the right side is given by a convection boundary condition, i.e.:

$$\vec{q}'' \cdot \vec{n} = h(T - T_\infty)$$

where h is a parameter known as the heat transfer coefficient. Obtain an expression for the nondimensional steady state temperature in terms of a nondimensional parameter known as the Biot number (Bi), given by: $Bi = \frac{hL}{k}$.

6.4   The book provided the boundary layer equations in scaled form. Rescale the boundary layer equations to dimensional form.

6.5   Scale Cauchy's momentum equation.

6.6   Consider Couette flow between two plates a distance of 2 millimeters apart whose top plate moves with a velocity of 1 m/s and the bottom plate is held fixed. The temperature of both plates are held fixed at zero degrees Celsius. The thermal conductivity is 0.6 $\frac{W}{mK}$ and the dynamic viscosity is 10^{-3} Pa · s. Find a value for the nondimensional temperature in the middle in the flow. Unscale the result and obtain a value for the temperature in the middle of the flow in Celsius.

6.7 Consider flow past a flat plate. At what point downstream from the leading edge does the boundary layer double in size compared to the size of the boundary layer at $x = 1$ meter? You may assume the Reynolds number never reaches 10^5, which is the transition to turbulence.

Further Reading

Anderson, John David, Degrez, Gérard, Dick, Erik, and Grundmann, Roger. 2013. *Computational Fluid Dynamics: An Introduction*. Springer Science & Business Media. This is a great beginner book to the world of computational fluid dynamics. It also has a great chapter on the governing equations.

Aris, Rutherford. 2012. *Vectors, Tensors and the Basic Equations of Fluid Mechanics*. Courier Corporation. A well known book for those more mathematically inclined.

Batchelor, George K. 1967. *An Introduction to Fluid Dynamics*. Cambridge University Press, xviii. This book has a reputation of being the bible of fluid dynamics. Be warned, however, even though the book title implies it is an introductory text, it is dense and can be intimidating to those uninitiated.

Dantzig, Jonathan A., and Tucker, Charles L. 2001. *Modeling in Materials Processing*. Cambridge University Press. While this book does not have fluid mechanics in the title, it covers the basics of fluid mechanics very well. In addition, I have found it to be one of the best texts covering nondimensionalization.

Fleisch, Daniel A. 2011. *A Student's Guide to Vectors and Tensors*. Cambridge University Press. One of the best introductory texts on vectors and tensors.

Munson, Bruce Roy, Okiishi, Theodore Hisao, Huebsch, Wade W., and Rothmayer, Alric P. 2013. *Fluid Mechanics*. Wiley Singapore. This is a beginning fluid mechanics textbook for engineering students. It covers the basics of fluid mechanics in general and has a considerable number of examples throughout the book.

Panton, Ronald L. 2013. *Incompressible Flow*. John Wiley & Sons. This book is a graduate textbook on incompressible flows. It covers a lot of ground, including a decent amount on compressible flows.

Warsi, Zahir U.A. 2005. *Fluid Dynamics: Theoretical and Computational Approaches*. CRC Press. This is an advanced textbook. It is one of the few books that fully covers the governing equations in non-orthogonal coordinate systems.

White, Frank M., and Majdalani, Joseph. 2006. *Viscous Fluid Flow*. Vol. 3. McGraw-Hill New York. This book has been used for many decades by engineering students studying viscous flows.

Index

Printed in the United States
by Baker & Taylor Publisher Services

Printed in the United States
by Baker & Taylor Publisher Services